ESSENTIALS of
TECHNICAL WRITING

William D. Conway

Ricks College

MACMILLAN PUBLISHING COMPANY

NEW YORK

Collier Macmillan Publishers

LONDON

Macmillan Publishing Company
866 Third Avenue, New York, New York 10022

Collier Macmillan Canada, Inc.

Library of Congress Cataloging-in-Publication Data

Conway, William D.
    Essentials of technical writing.

    1. English language—Rhetoric.   2. English language—
Technical English.   3. Technical writing.   I. Title.
PE1475.C66   1987   808'.0666   86–8729
ISBN 0–02–324560–3

Printing: 1 2 3 4 5 6 7   Year: 7 8 9 0 1 2 3

ACKNOWLEDGMENTS

Alyeska Pipeline Service Company, COMPANY DOCUMENTS. Reprinted by permission of Alyeska Pipeline Service Company.
ARCO Alaska, COMPANY DOCUMENTS. Reprinted by permission of ARCO Alaska.
Rose E. Jones Brown. BROWN. "The Swingline Tot Stapler." Reprinted by permission of Rose E. Jones Brown.
Leonard Clayton Dahle. DAHLE, "Report on the Feasibility of Wind-Generated Power." Reprinted by permission of Leonard Clayton Dahle.
Gould AMI Semiconductors, COMPANY DOCUMENTS. Reprinted by permission of Gould AMI Semiconductors.
Al Jephson, JEPHSON, "Replacing a Blown Fuse." Reprinted by permission of Al Jephson.
Tom Knudsen, KNUDSEN, Model, Classification: Arrowheads. Reprinted by permission of Tom Knudsen.
Jeanie C. Murphy, MURPHY, Model Letter of Application: Semi-Block. Reprinted by permission of Jeanie C. Murphy.
Darwin Reed, REED, Model 1, The Feasibility of Converting a Dairy Operation to a Fully Automated System. Reprinted by permission of Darwin Reed.
Stephen S. Reed, REED, Figures 2.3, 10.8, and 10.10. Reprinted by permission of Stephen S. Reed.
Jeffrey S. Stermer, STERMER, Model, Choosing a 35mm Camera. Reprinted by permission of Jeffrey S. Stermer.

ISBN 0-02-324560-3

TO BARBARA,
with love and appreciation

PREFACE

The title of this book suggests one of its main features; it presents the essentials of technical and professional writing. It makes no pretense to cover every specific writing situation that a graduate may encounter on the job. Rather, it is a concise introduction to essential writing techniques and structures, supplemented with models of student and professional writing, a broad selection of exercises and writing assignments, and a major section on editing technical prose. My goal has been to write a text that college students of all levels can understand, relate to, and even enjoy reading.

I assume that readers will represent a wide range of subject specialties and that some will have limited experience in writing. I focus on practical subjects that relate to readers' present experiences, backgrounds, and subject majors so that the problems that "inform" their writing are real to them. Accordingly, I include many examples of student writing. To broaden this experience, I place the discussion against a background of examples and references to business, science, and technology.

The book contains fifteen chapters grouped in five parts. Each chapter features review questions and an extensive selection of exercises.

The first part, "Getting Started," includes four chapters that define technical writing, explain principles of organization, emphasize introductions, and show students how to use effective visual support for their writing.

The second part, "Writing Basic Patterns," includes five chapters that discuss definition, description of a mechanism, classification and division, process description, and interpretation of data. The exercises include eight major writing assignments.

The third part emphasizes "Writing Business Correspondence" in two major chapters on business letters and memos. The discussion and examples are supported by many exercises based on examples from business and industry.

Because many courses in technical writing, as well as general college courses, require a major research paper, the fourth part, "Writing Research Papers," is particularly useful to students. This part also offers valuable background for writing proposals, research reports, and critiques on the job. Chapter 12 is a request for proposal (RFP) for a student research paper. Chapter 13 explains basic background, such as documentation and scientific attitude, and then discusses format and writing the paper; the discussion features a model student research paper. Chapter 14 shows readers how to write a critique—in this case, a critique of a research paper.

The last part, "Editing Technical Prose," a major feature of the book, contains sentences for students to revise. In Chapter 15, students can learn to master eight rules for gaining economy in their writing and then can gain practice by revising exercises based on student and industrial examples. Emphasis, concreteness, clarity, and correctness are other major topics covered in this chapter. These sections discuss wordiness, jargon, careless modifiers, careless pronouns, sexual bias, subject/verb agreement, fragments, parallel elements, and other essential principles. Every point is explained, illustrated with examples, and supported by exercises.

I have tested the manuscript with over four hundred college students in eight semesters and revised it three times. My colleagues have tested it in their evening continuing education classes. My favorite student comment concerning the book went something like this: "This is the first English text that I've understood and actually enjoyed reading."

Acknowledgments

Many people have helped me in developing the manuscript. John S. Harris of Brigham Young University was particularly helpful in sharing descriptions of his writing assignments. Ronald K. Messer of Ricks College provided much of the inspiration for the section on editing technical prose. My colleagues Donald Hammar and Eugene Thompson tested the manuscript in their classes and gave me encouragement to continue the project. Seth Bills, Chairman of Ricks College Educational Media, helped obtain the photographs.

My thanks to the students in my technical writing classes for their comments, criticisms, papers, and patience as the manuscript took form during four years of writing and testing. Particular thanks go to the following students who graciously gave me permission to use their writing as examples: Rose E. Jones Brown, Leonard Clayton Dahle, Al Jephson, Tom S. Knudsen, Jeanie C. Murphy, Darwin Reed, Stephen S. Reed, and Jeffrey S. Stermer.

My thanks to ARCO Alaska, Gould-AMI, and the Alyeska Pipeline Service Company for permission to use company documents as examples.

Finally, my thanks to my reviewers whose comments and insights were invaluable: Russell Briggs, Kalamazoo Community College; Donald H. Cunningham, Texas Tech University; Robert Gentry, TSTI–Waco; Blair Spencer Ray, Polk Community College; Marion K. Smith, Brigham Young University; W. Keats Sparrow, East Carolina University; and Thomas L. Warren, Oklahoma State University.

W.D.C.

CONTENTS

PART II
WRITING BASIC PATTERNS 71

PART **III**

WRITING BUSINESS CORRESPONDENCE   **141**

PART IV
WRITING RESEARCH PAPERS   187

PART **V**
EDITING TECHNICAL PROSE   **239**

PART I

GETTING STARTED

1

Technical Writing Defined

Written language is unique to the human race. We write and read plays, poems, songs, stories, letters, reports, instruction manuals, memoranda, and many other documents. Language is our way of sharing the past, present, and future. Yet not all uses achieve communication in the same way: an engineering proposal and a poem share similarities but have major differences in purpose.

This short chapter defines *technical* writing, identifies its general characteristics, explains how it differs from *literary* writing, and clarifies its importance in a technological society.

DEFINITION

Technical writing is written communication characteristic of business, science, and technology that emphasizes audience selection; precise, economical, unemotional language; certain organizational formats; and technical vocabulary.

Traditional technical writing territory includes material for business, science, and technology. However, the term has broader implications. Sewing, for example, has its own technical vocabulary—top stitch, blind stitch, even a "stitch in a ditch"—and detailed patterns, directions, and descriptions similar in complexity to those for adjusting or rebuilding a carburetor, using a piece of computer software, or preparing a specimen for microscopic examination. A book on physical conditioning or diet—a subject that may be treated casually or in depth—may have much in common with a manual for operating a small power plant. The point, then, is that even though technical writing is commonly associated with business, science, and technology, its features apply to any subject requiring careful, precise explanation, description, and direction.

Because clear written communication is so important in any profession, college students in such diverse fields as home economics, landscape gardening, welding, nursing, health, engineering, geology, physics, business, English, history, and so on, benefit from studying technical writing. The next section examines technical writing in greater detail.

GENERAL CHARACTERISTICS

Technical writing discusses or explains something in a factual, straightforward manner. The following are some of the more common features.

Concrete Language

Technical language favors the concrete rather than the abstract. Good writing uses a specific reference such as "B-52" jet bomber (see Figure 1.1) rather than a more abstract term like "weapons system." When abstract terms must be used, they should be supported by concrete detail.

> *Example:* "My purchase of word processing software (I bought Ezwrit) was the beginning of my love affair with the computer."

The information in the parentheses makes *word processing software* concrete.

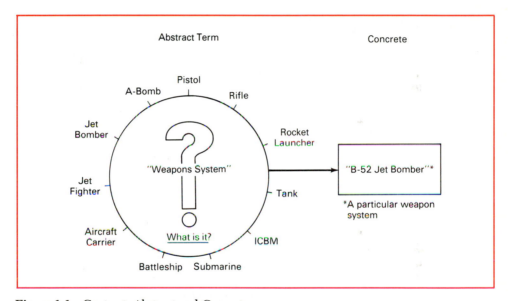

Figure 1.1   Contrast: Abstract and Concrete

Denotative Language

Technical language favors the *denotative* or objective, dictionary definition of a word. In contrast, *connotative* language refers to the subjective, abstract, emotional side of language. Compare the following:

DENOTATIVE	CONNOTATIVE
fifteen cents	inexpensive, cheap

The *fifteen cents* is concrete, factual, and, therefore, denotative. By contrast, the term *inexpensive* has a favorable connotation, meaning "low in price." The word *cheap* has a negative connotation, suggesting low quality as well as low price.

Objectivity

Technical writing strives for objectivity—an impartial, unemotional weighing of the evidence, sometimes called the *scientific point of view*.

Defined Audience

Because so many documents, reports, and memoranda are designated for small groups within a company or scientific organization, writers must tailor documents to meet the needs of these readers.

If audience analysis suggests that readers will know the technical language involved, an author needs to spend little effort defining and clarifying. If, however, some readers have other specialties and little knowledge of the subject, authors must define terms and give analogies.

Technical Style

Good technical writing is economical, clear, concise, and concrete. It gets to the point with little fanfare.

Negative Example: In the areas of technical writing and of business writing for college classes, most of the textbooks available are out of touch with the reality of the work place. (Many of these twenty-nine words are unnecessary. In particular, there are too many prepositional phrases.)

Revision: For business and technical writing, most college texts don't touch the reality of the work place.

Common Formats

Instruction manuals, proposals, technical reports, business letters, and memoranda use traditional plans or formats. To understand technical writing, you must recognize and use these common formats. The writing assignments in Part II of this text provide a thorough introduction to them.

Technical Vocabulary

Because technical writers deal with specialized subject matter, their writing is characterized by numbers, scientific symbols, specialized vocabulary, and acronymns—abbreviations made into words, such as NASA for National Aeronautics and Space Administration.

Example: It is obvious that the CKDR cell speed requirements are determined by the hold time requirements. Furthermore, comparing the above results with the CKDR capabilities (Table VI), we see that a CKDRL3 is required to adequately drive the shift register clock lines.

This passage is written for an expert audience, apparently, because it assumes knowledge of the terms *CKDR, cell speed, hold time, CKDRL3,* and *shift register clock lines*—all technical vocabulary.

COMPARISON—TECHNICAL AND LITERARY WRITING

Many characteristics of technical writing fit other types of writing as well. Certainly good writing, no matter what the application, is concise, clear, well-organized, and focused. However, technical writing and literary writing—poetry, drama, short stories, novels—differ fundamentally in purpose.

Technical writing is practical or informative. Literary writing, while providing some information, aims to give readers experience. Both are worthy goals! The following stanza from a poem clarifies this distinction.

The Fan

Scavengers had done their work well;
A lean frame, axles, treadless tires,
Transmission, and engine block remained,
Marked by the crossed blades of the fan.

In the leafless shadow of a gaunt tree
The skeleton of a car lay,
A luckless desert wanderer,
Half buried in the narrow shade.

. . .

WMC

The poet wants us to experience the scene. He helps us to see it and to compare in our minds the wreckage of the car to the bones of an animal or person. *Scavengers* has a double meaning—people who have removed parts, and flesh-eating birds like buzzards; *skeleton* reinforces the image of bones; *crossed blades of the fan* suggests a cross on a grave. Poetic language often has several levels of meaning.

The technical writer, by contrast, tries to deal with facts, with words that have just one meaning. This kind of writer's purpose is to inform. Dealing with the same scene, a technical writer would probably say something like this:

The chassis of a 1948 Ford, probably a sedan, complete with manual transmission (3-speed) and V-8 engine, motor #1289458, is located on a sand dune 25 miles southwest of XXXXXXXX, Texas.

Chassis, the proper technical term, has none of the emotional appeal of *skeleton*, making it appropriate for this writing situation. Facts—1948 Ford, manual transmission (3-speed), V–8 engine, motor number, location—give the reader specific information.

TECHNICAL WRITING AND MODERN SOCIETY

In a society characterized by new theories, discoveries, techniques, and products, precise speech and written communication play significant roles. For example, medical researchers around the world must understand what other researchers are doing. Because the knowledge explosion is so pervasive, the only practical way for them to share the vast amounts of data and information produced daily is to write about their own discoveries and to read about what others have done.

This explosion of knowledge causes nearly every profession to change dramatically every few years. Thirty years ago an electronics technician worked primarily with tubes and was adjusting to transistors. Integrated circuits were still a dream. Today's technician, working with integrated circuits having as many as one million transistors on a chip one-fourth the size of a finger nail, is dealing with problems that did not exist even five years ago.

New technology affects everyone, from the theoretical researcher to the consumer who must understand a product well enough to use it. Each process requires countless pages of technical communication. Indeed, technical communication is at the very heart of those activities that created the world we know and the one we will know in the future.

CONCLUSION

We all read and often must produce technical writing. Sewing, bow hunting, woodworking, performing lapidary skills, playing a musical instrument: all have their technical side, just as do medicine, scientific research, engineering, business, and other traditional technical fields. Technical writing is concrete, clear, impartial, audience-oriented. It contains traditional formats and a technical vocabulary, and differs from literary writing primarily in purpose.

REVIEW QUESTIONS

1. What is a definition of technical writing?
2. How does technical writing relate to fields outside of business, industry, and technology?
3. What is *concrete* language? Give an example.
4. What is the difference between *denotative* and *connotative* language? Which is favored in technical writing?
5. What is another term for *objectivity*?
6. What is *audience analysis*? Why is it important?
7. What are the characteristics of a good technical style?
8. How does *technical vocabulary* relate to audience analysis?
9. What is the basic difference between technical and literary writing?
10. How important is technical communication in our society?

EXERCISES

1. Write a specific, concrete form for each of the following abstract terms:
   a. car . . . . 1986 red hatchback (Example)
   b. tool
   c. book
   d. machine
   e. device
   f. business form
   g. music
   h. typewriter
   i. movie
   j. president

2. Choose the word that has the most positive connotations.
   a. traitor, patriot
   b. loyal worker, strike breaker
   c. peculiar, strange, odd
   d. confidential, secret, restricted
   e. confess, acknowledge, admit
   f. destroy, demolish, lay waste
   g. soil, loam, dirt
   h. portly, obese, fat
   i. offensive, gross, disgusting
   j. leave, desert, abandon

3. Writing may be subjective, objective, or it may combine the two qualities. Evaluate the following passages, placing one line under objective language and two lines under subjective language.

a. I believe this is the superior product because it is better than all the others.
b. In that situation, the President should have sent in the marines.
c. The 40 watt bulb costs 85 cents.
d. The technician used too much solder on the joint.
e. The material was soft and white in appearance.
f. The material is an 8 on Moh's scale of hardness.
g. The new machine weighs 22 tons.
h. People should wait until they are at least twenty-five to get married.
i. The contractor used 8-inch casing for the first 1,200 feet of the well.
j. The new machine looks like a pile of junk.

4. A skillful technical writer usually attempts to communicate with a particular audience. Based upon the length of the sentences, the complexity of the material, the difficulty of the technical language, the use of comparisons, and definition of terms for the following passages, identify the audience as *novice*, *intermediate*, or *advanced* in its understanding of the material discussed. Explain what evidence led to your selection.

a. Accurate navigation is easier than ever because now you can get all the quality and reliability of the popular Loran C Navigator with even more waypoints.
b. A high-efficiency filter removes 99.97 percent of particles 0.3 microns or larger in size (a human hair is about 50 microns in diameter).
c. The new device, about the size of a D battery, is a sophisticated pump designed to assist the human heart.
d. Although the new scrubber looks pretty complicated, it is really nothing more than a fancy washing machine.
e. The wind is blowing. It is cold. It is time to build a fire to keep warm.
f. Each black pixel has the binary value 00. Therefore, each corresponding screen mask pixel (2 bits) must contain the value 00, and each corresponding cursor mask pixel must contain the value 00.

5. Here is the rest of the poem quoted on page 7. Translate it into technical English.

The wind came; sand rose and walked across
The dunes in smooth flowing ripples.
The fan swung in a tentative circle—
Each blade individually clear, then blurred,
like a plate suspended on its side in the air.

The wind cascaded down a miniature valley
Whirled up a hill and disappeared.
Blades reappeared slowly, clearly.
A last slow arc scored the sky, slipped back, stopped.

Up the next hill I trugged;
I smiled when I looked back.
The skeleton was smaller now,
A silent, steel relic—with a pinwheel on its nose.

2

Basic Principles

Technical writing, like any discipline, has basic principles that must be mastered before advanced concepts can be tackled. To help you get started, this chapter concentrates on five of the most important guidelines:

1. Use three-part organization
2. Use emphatic structure
3. Analyze the audience
4. Write a thesis statement or statement of purpose
5. Use an outline.

USE THREE-PART ORGANIZATION

Three-part organization is of *extreme* importance in technical writing. A common saying among business and technical writers goes something like this:

- First, you tell them what you're going to tell them.
- Then you tell them.
- Then you tell them what you've told them.

Figure 2.1 shows the relationships among several commonly used terms, all of which are associated with three-part organization.

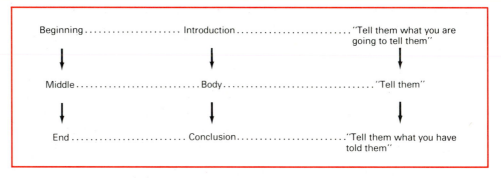

Figure 2.1   Three-Part Organization: Related Ideas

Three-part organization may also be compared to a journey. Just as a traveler needs to know where to start, which way to go, and where to go, so, too, does a reader need directions in the introduction to a document. The forks in the road and the sights are similar to the paragraphs and divisions in a piece of writing: both types of information need to be signaled clearly so that no one becomes lost or confused. Looking back to remind the traveler of direction

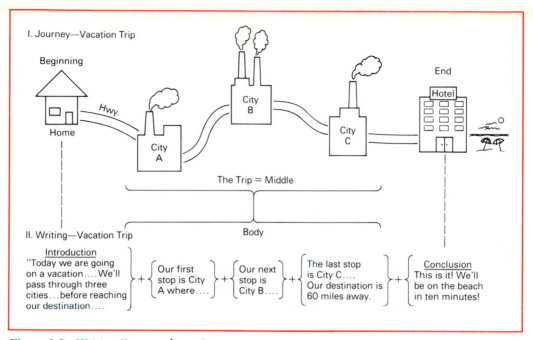

Figure 2.2 Writing Compared to a Journey

and position is a process similar to writing a transitional statement that says, in essence, "So far we've discussed this, and now we're going to inform you about another aspect." Just as a traveler needs to know when to get off the bus, so does the reader need a clear signal showing when a document ends.

Figure 2.3 is the introduction to a formal report—the part that "tells them what you're going to tell them." Notice how it gives an effective overview of the subject, purpose, scope, and statement of organization.

The first paragraph of Figure 2.3 gives general background and clarifies the *subject* and *problem*—jack rabbits and how to control them. The second paragraph establishes the *purpose* for writing, *criteria* for evaluation (cost, efficiency, and humane measures) and then tells *how* the document is *organized*. This approach gives the reader an overview of what to expect in the document.

Three-part organization is also common in paragraphs of technical writing. The introduction is represented by the topic sentence; the body is equivalent to the discussion that follows; and the conclusion is stated in the last sentence.

Figure 2.4 contains a paragraph split into a beginning, middle, and end to clarify the position and relationship of the three parts.

No matter what you write, keep three-part organization in mind. Figure 2.5 shows the universality of this familiar pattern.

INTRODUCTION

Since the first crops were planted until the present, western farmers have been faced with crops damaged by jack rabbits. This animal's ability to reproduce at an incredible rate, along with its love for farm vegetation, has brought it nation-wide attention. Though farmers have been losing their crops for over a century, controversy has always been present over the means of controlling the jack rabbit population.

The purpose of this report is to determine if there is an "acceptable" method to control rabbit population in Southeastern Idaho. Criteria are cost, efficiency, and humane measures. The discussion includes a description of the rabbit and problem; comparison and analysis of biological, mechanical and chemical means of control; summary and recommendations.

Figure 2.3 Introduction to a Formal Research Report

A synthetic woven fabric provides a framework for casting in place cable-tied concrete blocks.
                              For this approach, on-site batching facilities would be required to supply a pumpable sand/cement mortar. The concrete used in this system would be less durable than the concrete of the articulated block system in its ability to resist abrasion and deterioration due to freeze-thaw cycles. This system is a new product with only four months of manufacturing and marketing history.
                              Due to performance un-certainties, the risk of using this product is higher than that with the articulated block system.

Figure 2.4 Sample: Paragraph Structure

USE EMPHATIC STRUCTURE

Many writers are surprised to learn that their writing has emphasis—whether they intended it to or not! Simply placing material at the beginning or the end (or both) of a document gives that material emphasis. The beginning or introduction creates the reader's first impression of the document; the conclusion presents the last.

Failure to understand the principle of emphasis causes communication problems. One wordy supervisor thought his employees couldn't read very well

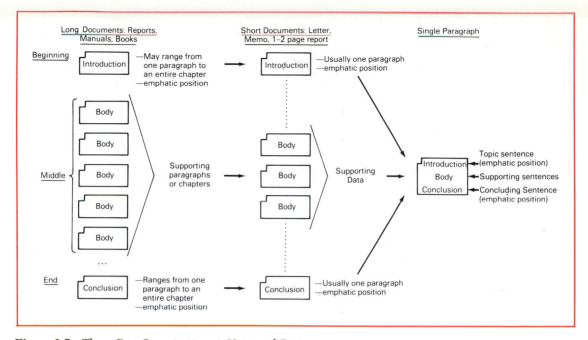

Figure 2.5   Three-Part Organization: A Universal Pattern

because they never seemed to understand *his* memos. In the first three or four lines of each memo, he dealt with trivial points until he warmed up to his subject. After discussing the main point in the body of the memo, he usually concluded by adding several afterthoughts. His readers typically skimmed the first few lines for main ideas, skipped the body entirely because of his wordy writing style, and then checked the last few lines to see if they had missed any key points. The result was, naturally, poor communication.

As you write and revise your thoughts, consider this statement to contain the basic rule of emphasis:

> Sentences, paragraphs, and longer documents have a beginning, middle, and end. The beginning and ending positions are the most emphatic and the middle is the least.

If you use three-part organization consistently, you will automatically place key ideas in emphatic positions. As you revise documents, particularly those that are confusing and difficult to understand, you should place key ideas in the first and last slots in paragraphs, chapters, sections, and complete documents. Careful writers even apply the principle of emphasis to the shortest of sentences.

Figure 2.6 is a letter sent by a large construction company to a smaller firm seeking a bid for the design, fabrication, and erection of a metal building.

This is an RFP(?)

What is emphasis?

Dear Mr. Maxwell:

   XXX Mechanical Division is currently developing a proposal to perform construction of an air separation plant in
XXX County, Iowa. Included in the scope of our bid is a steel
building intended to house large compressor units. The dimensions of this structure are 360 feet long by 110 feet wide by
68 feet high. Also included is a 50 ton overhead crane which
shall be installed at a height of 60 feet and run the entire
length of the building. I am enclosing what little information
I have including site conditions and a foundation print. I
would appreciate receiving a proposal for the supply and
erection of the building by May 22, 1986. As more information
becomes available I will forward it to you, although the specifications do indicate that the contractor is responsible for
design.
   My telephone number is 208-356-xxxx. Any questions you
have should be called in as soon as possible due to our tight
bid schedule.

Figure 2.6   Negative Example: Organization

The larger company needed a bid for its own proposal to develop an entire
plant site. Evaluate the letter for three-part organization and emphasis.

   The author of this letter was apparently not aware of three-part organization
and emphatic order. If Mr. Maxwell was not discouraged by the length of the
first paragraph, he probably kept thinking, "Why are you writing me?" and
"What do you want?" Normally these questions are answered in an introduction.

   Just past the middle of the letter, the author finally explains his purpose:
"I would appreciate receiving a proposal for the supply and erection of the
building by May 22, 1986." At this point Maxwell probably stopped reading
and shifted back to the beginning, because he now understood the significance
of the detailed information.

   After re-reading the detail concerning the building, Maxwell probably read
the last paragraph (the conclusion) that says the bid schedule is tight. The
author, however, failed to list a deadline. The date, of course, is buried in the
middle of the letter—in the least emphatic location.

   Figure 2.7 contains a revision of the opening and closing paragraphs of
this letter. Key data have been moved to these locations. Which version do
you prefer?

   The three-part organization and emphatic order of the revision are effective
because they call attention to key data in the introduction and stress the due
date in the conclusion.

   Closely related to emphasis is the *order of ideas*. Writing can be "inductive"

Dear Mr. Maxwell:

As part of a proposal to construct an air separation plant, XXX Mechanical Division would like to invite you to bid on the design, supply, and erection of a steel building.

[The body of the letter would, logically, contain all the specifications currently available on the "steel building."]

Because we are on a tight bid schedule, please have your bid to us by May 22, 1986. If you have questions, call 208-356-XXXX.

Figure 2.7   Revision: Sound Organization

or "deductive." Nearly all business and technical writing is deductive: that is, the main idea or direction of the paper appears at the beginning. Inductive order, which doesn't reveal the main idea until the end of the paper, is uncommon.

Inductive order is occasionally useful if a reader may be offended by reading the key idea in the opening. In such a circumstance, the writer may want to reveal some compelling support for the main point before revealing it. Generally, however, technical writing emphasizes deduction—letting the reader know the main idea from the start.

The role of emphasis in organization is particularly important when a writer designs a document. The following section explores five patterns of organization.

Order of Time

Application:	Accident reports, day-to-day construction reports, factory procedures or processes, histories of a company or individual, detailed instructions.
Advantages:	Easy to keep coherent. Time words make good connectors: one year later, the next day, the following step, next, before, and so on.
Disadvantages:	Does not provide proper emphasis. Often buries key data that a busy reader wants to find quickly and easily.

Order of Place

Application:	Reports organized by divisions of a company, sales reports organized by geographical areas, details organized from top to bottom, near to distant, and so on.
Advantages:	Shares many advantages with order of time. Spatial words connect sentences and paragraphs: i.e., *where, farther left, on top*. Easy to write.

| Disadvantages: | Does not stress points of interest or by itself hold reader's interest. Monotonous in long papers. Writer still has problem of beginning. |

Order of Increasing Emphasis

Application:	Engineering reports, research reports, and many other common uses. The following pattern was once considered the best order for reports on experimental work: • statement of problem • description of equipment • discussion of procedure • statement of results • discussion of results • conclusions and recommendations.
Advantages:	Has gradually increasing interest. This order is traditional in science and technology. Some readers find it familiar—and therefore safe.
Disadvantages:	Many readers are in a hurry and want main ideas quickly.

Order of Decreasing Importance

Application:	Engineering reports, research reports, and many other common uses. Many readers didn't like order of increasing importance, so government and business reversed the order to place key data first.
Advantages:	Places key ideas first, makes strong initial impression, begins with material of interest to the reader.
Disadvantages:	Has decreasing reader interest. Some readers lack sufficient knowledge to understand conclusions and recommendations before reading the report.

Order of Emphasis by Position

| Application: | Engineering reports, research reports, and many other common applications. Writers sought a plan to eliminate the undesirable features of increasing and decreasing importance and preserve their strengths by using a short summary of main points prefixed to a report organized in the order of increasing importance. |
| Advantages: | Has high reader interest. A busy reader can read a few sentences of a report, letter, or memo, and be confident that those lines contain all information needed to judge whether to read the rest of the document. It uses the positions of emphasis to a maximum. |

Disadvantages: Can become a monotonous formula. Some material should be introduced inductively to avoid antagonizing a reader.

Each of the five patterns works to communicate information. Practice varies among writers according to personal preference and purpose in writing. Order of time and order of place tend to conceal key data. Nevertheless, if these two schemes are modified to stress key ideas in emphatic positions, they can be effective approaches to some subjects. Each of the last three patterns has supporters. This text favors the last because it stresses three-part organization as well as emphatic order.

ANALYZE THE AUDIENCE

There are several steps to follow before you write a letter, memo, or report, or give a talk. As part of your prewriting preparation, analyze the audience by asking yourself these questions:

- Who is the primary reader?
- What is this reader likely to know about the subject?
- What is the reader's attitude toward the subject?
- Who else will read what I have written?
- What should the reader know after reading?
- What attitude should the reader have after reading?

In *Audience Analysis for Technical Writing*, Thomas E. Pearsall identified five audiences with varying needs: the layman, the executive, the expert, the technician, and the operator. Each audience has differing needs:

- *Layman* needs help to understand the subject, including definitions and analogies.
- *Executive*, who makes decisions involving the entire company, wants an overview.
- *Expert* understands the subject and wants a good technical discussion.
- *Technician* wants a little theory with specific detail about how to repair or build something.
- *Operator* wants to know how to use something.

Your job as a writer is to meet the needs of the reader. If you write a memo asking someone to give a talk, for example, but fail to say when the talk is to be given, the reader will have to write or call you. Do the job right the first time—by meeting the reader's needs.

You may be wondering what to do if several audiences read the material—

as will often be the case. In general, aim at the lowest level of understanding. You risk boring the expert but all audiences will be able to follow your points. If you cannot define the audience clearly, make the best estimate you can, and then write for that audience.

Analyze the audience in Figure 2.8. Which of the five audiences do you think it fits—layman, executive, expert, technician, or operator?

---

1. Provide IRIG B formatted time to be a time base for analog instrumentation tape recorders unmodulated and 1 KHz sine wave modulated.
2. Provide paralleled BCD output, 42 bits, TTL compatible for driving remote displays or a tape search controller.

Figure 2.8

What is your reaction to Figure 2.8? Unless you are knowledgeable about this particular subject, words like *IRIG B, formatted time, time base, analog instrumentation, unmodulated,* and *1 KHz sine wave modulated* are probably confusing. If the reader of this passage is a technician or an expert, these terms are not confusing. But if the audience is a layman or executive, the author failed to communicate!

If your audience cannot understand what you are attempting to say, there is little point in writing. A novice or layman may require background, definitions, a simplified statement of theory, and a gradual introduction to technical vocabulary. Most writers err by assuming that readers know more than they really do. This error is easy to make if you work with a subject daily—you gradually forget the struggle you had in mastering it.

WRITE A THESIS STATEMENT OR STATEMENT OF PURPOSE

Closely related to three-part structure and emphasis are the thesis statement and statement of purpose. Both are ways to introduce succinctly the main idea in one sentence at the beginning of a document. Writing these statements is an excellent way for you, as a writer, to recognize and to focus your ideas before you attempt to write a document.

As you formulate a thesis idea and/or statement of purpose, you should ask yourself these two questions: Why am I writing? Who will read the memo?

Figure 2.9

In business and technology nearly every written document arises from a need within an organization. *Purpose* and *need* are key factors to consider.

A thesis statement is a declarative sentence that states an opinion about the main idea. Here is an example:

Example: The XXX-XXXX microcomputer meets all company specifications for memory, speed, and ease of operation.

The example states the author's opinion that the computer "meets" company specifications. As a thesis statement it also suggests the topics for the main sections of the paper—specifications, memory, speed, and ease of operation. The author wrote the statement in this form as a response to a need in the organization, apparently to evaluate the computer as a possible purchase.

The thesis statement not only helps the reader but also helps you as an author focus on your topic. If you cannot write your main idea in one sentence, you probably do not have a clear idea of what you want to say. Do a little more research, make lists of some ideas to include, reassess carefully why you are writing and who will read it. Then try again. Once you have captured the main idea in a sentence, writing the document will be much easier.

A thesis statement serves as a filter of ideas. As you think about all the data you have available, the thesis statement helps you recognize which parts relate to your topic. Study the following thesis:

Example: Process #1 produces a quality chip more reliably than Process #2 and, therefore, should be adopted for production.

Notice how this thesis leads the writer to appropriate supporting material.

1. What was Process "#1? Process #2?
2. What quality standards were used?
3. What is the difference in reliability?
4. What is the cost difference?

Answering such questions provides most of the information needed to develop the body of the paper. A thesis helps you focus the subject and limit the supporting concepts to a number appropriate for developing a particular subject.

A *statement of purpose* is closely related to the thesis statement, but lacks the element of personal opinion. Rewritten as a statement of purpose, the thesis statement on chip manufacturing could now read this way:

Example: The purpose of this paper is to compare Process #1 and #2 and recommend one as a manufacturing procedure.

A statement of purpose doesn't take a stand (it does not hold that one product is superior); it simply states the situation. Both forms are helpful to the reader—and writer—because they announce key ideas in an emphatic location, the introduction.

USE AN OUTLINE

Because clear organization provides the foundation for successful technical writing, using an outline can be extremely helpful as a prewriting activity. This is especially true when the writer also assesses his or her audience and produces preliminary attempts to write a thesis statement or statement of purpose. Few writers have the ability to organize their thoughts entirely in their heads and then keep track of them through several interruptions.

There are three types of outline: the topic, sentence, and paragraph:

- *Topic Outline*   words or phrases only
- *Sentence outline*   sentences only
- *Paragraph Outline* paragraphs only

Topic and sentence outlines are widely used in technical writing; the paragraph outline is a rarity. This section emphasizes topic outlines.

There are five steps to writing and using an outline:

1. Write a thesis or statement of purpose.
2. List all topics and ideas.
3. Group the topics.
4. Test the outline.
5. Use the outline to write.

Write a Thesis or Statement of Purpose

As was pointed out earlier, writing a thesis statement or statement of purpose helps you find and focus your subject. Notice how beneficial the following example is to the author.

Example: The purpose of this report is to differentiate between procaryotic and eucaryotic cells by discussing the principal features of each.

A paper following this statement of purpose will probably define two types of cells; then it will list and discuss the principal features of each. These are the very ideas required for the outline!

List All Topics

A good tactic in outline writing is to list on the left side of several sheets of paper every idea you can think of that may be worth including. Do not worry if some are major ideas and some are less important; you will straighten them out later. Figure 2.10 provides an example of how your list may appear.

If you have had difficulty writing a thesis or purpose statement, listing ideas is an excellent way to focus your attention.

Group the Topics

Now study your list and attempt to group those ideas that seem related in some way. Use the right side of the page. If some topics are not important, cross them off. Using your thesis or purpose idea as a guide, you will begin to notice relationships among the items.

Once you have completed your second column, study the groups of ideas and make a tentative guess about a logical order for presenting them. Indicate your choice by numbering the topics in a logical sequence. Study Figure 2.10: the first column is the original list of ideas, the second column is the first attempt at grouping ideas.

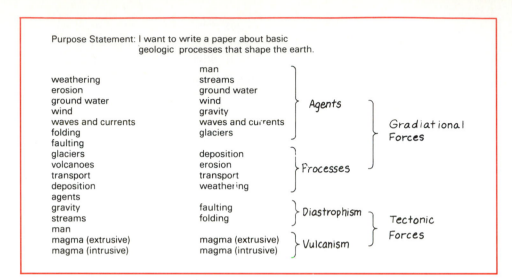

Figure 2.10   Listing and Basic Grouping

Write an Outline

Listing topics, grouping them, and writing an outline tend to be one continuous process. Figure 2.10 illustrates the listing and tentative grouping process. The handwritten notes show the author's attempt to bring order to the ideas by grouping common themes together: processes, agents, diastrophism, and vulcanism. By now you can probably guess the structure of the outline that resulted. Study the outline finalized in Figure 2.11.

There are several ways to number outlines. Two common patterns are the number system used in this book and the Roman numeral pattern. The number system is widely used in business and technology. Figure 2.12 shows how both systems appear.

Before you actually use an outline to write, test it to see if it covers all aspects of the subject in logical divisions.

Test the Outline

A good outline should cover the subject adequately and break it into logical divisions.

To test your outline, see if you can add anything to it. For example, consider this section from the outline in Figure 2.11:

Example: 3.1 Processes
               3.1.1 Weathering
               3.1.2 Erosion
               3.1.3 Transport
               3.1.4 Deposition

```
1.0 Introduction

2.0 Tectonic Forces
    2.1 Diastrophism
        2.1.1 Folding
        2.1.2 Faulting
    2.2 Vulcanism
        2.2.1 Intrusive Magma
        2.2.2 Extrusive Magma

3.0 Gradational Forces
    3.1 Processes
        3.1.1 Weathering
        3.1.2 Erosion
        3.1.3 Transport
        3.1.4 Deposition
    3.2 Agents
        3.2.1 Streams
        3.2.2 Wind
        3.2.3 Glaciers
        3.2.4 Waves and Currents
        3.2.5 Gravity
        3.2.6 Ground Water
        3.2.7 Man

4.0 Conclusion
```

Figure 2.11   Sample Outline

```
NUMBER SYSTEM                          ROMAN
1.0                                    I.
    1.1                                    A.
    1.2                                    B.
2.0                                    II.
    2.1                                    A.
    2.2                                    B.
    2.3                                    C.
        2.3.1                                  1.
        2.3.2                                  2.
            2.3.2.1                                a.
            2.3.2.2                                b.
            2.3.2.3                                c.
```

Figure 2.12   Numbering Systems Compared

Weathering, erosion, transport, and deposition compose the main topic *Processes*. You can thus test the coverage of this part of the outline by trying to think of any other subject that is a gradual geologic process. If you can think of another process, add it to the outline. If not, then this section has complete coverage.

Once you have established the coverage, check the outline for logic. Data grouped on each level in one section of the outline should have a common basis of classification. Study the following example:

Example: 2.0 Types of Hulls
            2.1 Monohull
            2.2 Catamaran
            2.3 Trimaran
            2.4 Ketch
            2.5 Schooner

Although items 2.1–2.5 refer to boats in some way, the author used a mixed basis of classification. Items 2.1, 2.2, and 2.3 are logical entries under the main heading *Types of Hulls*. However, items 2.4 and 2.5 have a different basis of classification: they refer to the number and placement of the masts and sails on a hull, not to the number or shape of the hulls. In fact, the monohull, catamaran, and trimaran could all be rigged as ketches or schooners.

A revision might read this way:

Example: 2.0 Types of Hulls
            2.1 Monohull
            2.2 Catamaran
            2.3 Trimaran
         3.0 Sail and Rigging
            3.1 Ketch
            3.2 Schooner
            3.3 Sloop
            4.4 Yawl
            4.5 Cutter
          …   etc.

In the revision, each subsection of the outline has only one basis of classification: number of hulls in the first, and arrangement of sails in the second. Another way to correct the original is to remove *ketch* and *schooner*. If the author wanted to concentrate only on types of hulls, this would be an effective approach.

If an outline has adequate coverage and a logical basis of classification, so will a document based upon it.

Use the Outline to Write

Now that you have evaluated your outline for coverage and logic, you are ready to use it to write your paper. An outline provides a guide for you to write introductions, headings, transitions, and supporting material.

Figure 2.13
Basis: Key to Classification

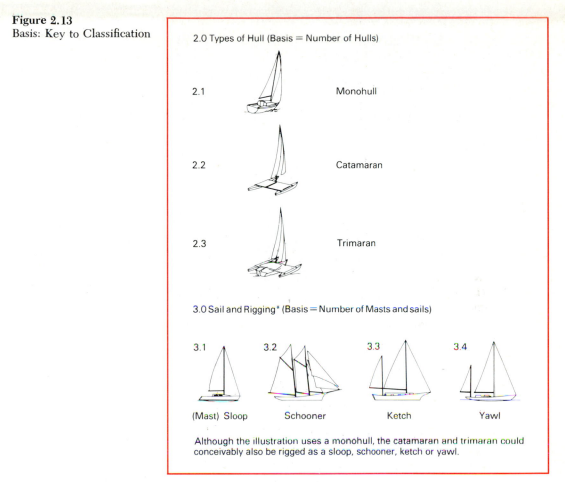

2.0 Types of Hull (Basis = Number of Hulls)

2.1   Monohull

2.2   Catamaran

2.3   Trimaran

3.0 Sail and Rigging* (Basis = Number of Masts and sails)

3.1   3.2   3.3   3.4

(Mast) Sloop   Schooner   Ketch   Yawl

Although the illustration uses a monohull, the catamaran and trimaran could conceivably also be rigged as a sloop, schooner, ketch or yawl.

The following example (Figure 2.14) is based upon the research report on rabbit control mentioned in Figure 2.3. It illustrates the relationship between the outline and the material the author included in the paper.
Comparison of the outline and writing sample shows how close the relationship is between them. Following is a step-by-step analysis of the two:

1. OUTLINE: "IV. BIOLOGICAL CONTROL"
The main heading from the outline tells the writer to do two things:

a. Write a main heading in the paper.
b. Write an introduction to the chapter.

The first paragraph in Part B of Figure 2.14 is the resulting introduction. In the first sentence the author says that this is but one of three sections of the

Example A: (From the outline for a paper on controlling rab-
             bits)

4.0 Biological Control   — Heady
    4.1 Predation
    4.2 Habitat Manipulation
    4.3 Diseases and Parasites
    4.4 Chemosterilants

Example B: (From the body of the paper based upon the outline)

IV. Biological Control   — Heading

Biological control is one of three types discussed in the body of this report. Though probably the most intricate approach to solving the jack rabbit problem, the biological means has only limited application in control. Predation, habitat manipulation, diseases and parasites, and chemosterilants are the four major types of biological control discussed in this section.

Predation

Predation is the use of natural enemies to control population numbers. . . .

Figure 2.14   Relationship: Outline to Written Text

paper: this is a transition to tie this section to the others. After pointing out that biological control has limitations, the author tells what to expect in the sections that follow: a discussion of four methods of control. Because the methods are listed in a particular order, readers expect them to be discussed in that order.

Analyzed another way, the introduction gives the following information:

- *Subject*—biological control
- *Purpose*—to discuss means of biological control
- *Scope*—four major types
- *Plan of Organization*—discuss four points in the listed sequence

2. OUTLINE: "SECTIONS 4.1–4.4"
Each subheading tells the author to do three things:

a. Write a heading.
b. Write a transition.
c. Write specific data concerning the heading.

It should come as no surprise to the reader that the first subheading in part B of Figure 2.14 is "4.1 Predation." This is what both the outline and introduction signaled would come first.

    The author tied his ideas together in the next section. After the heading "4.2 Habitat Manipulation," the first line said the following:

> *Example:*   The second method of biological control, habitat manipulation, means to. . . .

The first six words tie this section to the introduction: readers now know they are reading about the second of four types of control mentioned in the introduction.

    Although not included in Figure 2.14, the second, third, and fourth subheadings are easy to guess—from the outline and the introduction.

    The fourth subheading also tells the author to write a transition signaling the conclusion of this section. The wording could be something such as "The final control discussed in this section is a chemosterilants." Or the author could produce a concluding paragraph. The concluding paragraph could be a summary, a conclusion stressing the value of one of the four controls, or some other approach that clearly signals

"This is the end."

An outline, properly used, is an excellent way to structure writing for both you and the reader.

CONCLUSION

The five basic principles of technical writing introduced in this chapter are valuable background for all the writing assignments included in this text.

REVIEW QUESTIONS

1. What is three-part organization?
2. How does three-part organization relate to paragraphs?
3. What is emphatic structure?

4. How does emphatic structure relate to sentences, paragraphs, and longer documents?

5. What are inductive and deductive order?

6. Which form (inductive or deductive) is most common in business and technical writing?

7. Which of the five patterns of organization does this book recommend?

8. What are the strengths and weaknesses of the recommended organizational pattern?

9. Why is audience analysis important to both the reader and writer?

10. What are the five audiences identified by Pearsall?

11. How do the needs of these audiences differ?

12. As an author, for whom do you write when you have multiple audiences?

13. What is a thesis statement? What are its characteristics?

14. How is a thesis statement helpful to both reader and writer?

15. What is a statement of purpose? How does it differ from a thesis statement?

16. Which form of outline (topic, sentence, or paragraph) does this text stress?

17. How can you test an outline?

18. What is meant by *the logical basis of classification* for an outline? How do you check its logic?

19. How does an outline signal you to write introductions, headings, transitions, and supporting material?

20. How does an outline signal you to write a conclusion?

EXERCISES

1. Evaluate this chapter for three-part organization. Does it have a beginning, middle, and end? Are key ideas emphasized at the beginning and/or ending?

2. Evaluate the following memo for three-part organization and use of emphasis.
   a. Read the entire memo several times and then underline the paragraph that contains key information.
   b. Does this document use the positions of emphasis properly? What changes would improve the memo?
   c. What information would a busy reader want to find in the first paragraph? Which paragraphs contain supporting information?
   d. Based upon clues found in the following memo, identify the audience.

   Because of ACME's recent growth, the number of employees has increased significantly. As ACME grows, changes need to be made in procedures in order to operate effectively.

   Currently, all vacation requests are submitted to Accounting. Accounting has been responsible for monitoring requests to make sure the time is reported on the timecard. This control needs to be placed closer to the source. First level supervisors are better able to control the vacation time taken by their employees.

Vacation requests will be turned into the department and the department will be responsible for making sure vacation time taken is recorded on the timecard.

Accounting will help you track vacation time by providing your area with a weekly list of vacation time taken the previous week.

If you need any help in establishing a method for tracking vacation time, please give me a call.

3. In the college library compare several magazines such as *Scientific American, Road and Track,* and *Popular Science* for attention to audience.
   a. What do the advertisements indicate about the education and income level of the audience targeted by advertisers? Do the levels vary in any way among the three magazines? Explain. Do the types of products vary? Why?
   b. Evaluate the content of the periodicals. Which periodical do you find easiest to read? Why? Which of the three requires an audience with the highest level of education? What is the basis for your selection?
   c. Compare illustrations used as graphic support for articles in the three periodicals. Which uses the most complex illustrations? Which the simplest? What generalization relating to the three periodicals can you make about audience?

4. Write a thesis statement and then a purpose statement for each of the following situations:
   a. You work in the college cafeteria. You feel the way your supervisor schedules workers is unfair. You believe morale is low as a result. You feel a regular schedule with a rotation of shifts for evenings and weekends will help solve the problem. You decide to write a tactful memorandum stating your position.
   b. You work for a data processing firm. Your boss has asked you to evaluate three different software programs for inventory control. After testing each program, you do not feel comfortable recommending any of the three. Because of various problems, each program is inadequate for company needs. You decide to write a memo stating your position.

5. Select an article (approximately 1,200 words) from a journal in your field and write a topic outline of the article. Is the article logically organized? Explain why or why not.

6. Analyze the following outline segments for logic. Remember each level should have just one basis of classification.
   a.   2.0 Types of Outlines
            2.1 Topic Outline
            2.2 Sentence Outline
            2.3 Paragraph Outline
            2.4 Document Outline

b.   4.0 Mechanical Means of Control
       4.1 Traps
       4.2 Barriers
       4.3 Guns
       4.4 Predators

c.   3.0 Technical Duties of a Clerk-Typist
       3.1 To operate office equipment efficiently
          3.1.1 Typewriters
          3.1.2 Photocopy machines
          3.1.3 Adding machines
          3.1.4 Word processing machines
          3.1.5 Other
       3.2 To answer the telephone courteously
       3.3 To be on time
       3.4 To keep the files
       3.5 To maintain an attractive appearance

3

Introductions

When readers pick up a report, memo, letter, technical journal article—almost any written work—they need information:

1. What are you writing about?
2. Why are you writing about it?
3. How does this subject relate to me?
4. What can I expect to find if I read on?
5. In what order will you present ideas?
6. What conclusions or recommendations do you make?

Your job as an author is to include some or all of this information in the introduction. This way, you provide a framework on which to hang the ideas that follow. Of course, the content and form of an introduction depend on the problem that creates your need to write, on your purpose, on the demands of the subject, and on your analysis of the reader's needs.

This chapter builds on the concepts presented in Chapter 2. The first section discusses identifying the problem; the second explains the four basic elements of an introduction; the third explains how to divide or classify business and technical writing topics into specific sections that signal the organization of the body of a paper; and the final section explains three common variations of the summary beginning.

IDENTIFY THE PROBLEM

Writing doesn't occur in a vacuum; something creates the need that encourages or forces a writer to place pen on paper. Most documents are written in response to a problem or situation. Recognizing the problem—that is, finding the reason you must write—is fundamental to your efforts to get started on a document.

In college, for example, an assignment to research and write a 4,000-word research paper is a *problem* or *writing situation*. Your response, of course, will be to write the paper. In business, science, industry, and other fields, writing results from an author's need to solve a problem.

Figure 3.1 illustrates how a particular problem in a manufacturing company not only can hinder or stop production but can also generate a series of written documents.

Hundreds of similar problems occur daily in business and industry. A minor misunderstanding of an employee concerning vacation benefits may require a single memorandum to solve the problem. The planning and building of a new plant, in contrast, may require years of planning and countless letters, memoranda, reports, diagrams, schematic drawings, and other documents.

Recognizing the problem that creates or *informs* a writing situation is an important step in prewriting because the problem, in some way, is the subject of the document that results. This recognition is basic to writing the four basic elements of an introduction discussed in the next section.

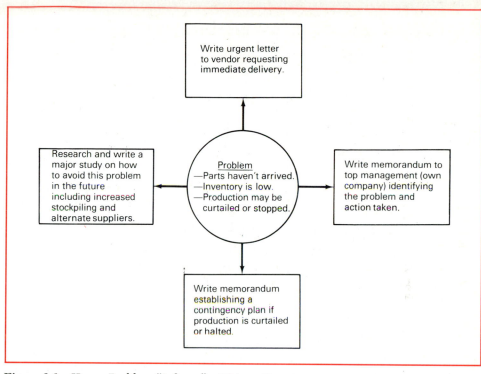

Figure 3.1   How a Problem "Informs" a Written Document

USE FOUR BASIC ELEMENTS

Introductions are amazingly similar in content: most include four statements:

1. Subject or problem
2. Purpose
3. Scope
4. Organization

In addition, some introductions contain definition, theory, historical background, or other information helpful to the intended audience. Figure 3.2 shows the relationship between the four statements and the interests of the reader.

Subject

Readers first want to be informed about the subject or problem. Sometimes the title makes the subject clear. However, the introduction should state it also. Study Figure 3.3.

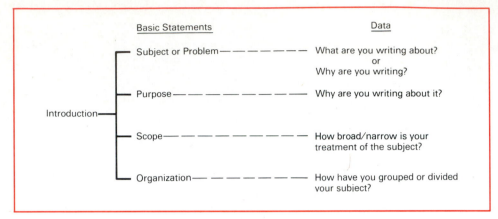

Figure 3.2   Introduction: Basic Form and Content

CELLS: A STRUCTURAL ANALYSIS
The fundamental unit of all living systems is the cell. A major
characteristic that distinguishes cells from one another is
internal structure. On this basis, cells can be placed into
two categories: procaryotic or eucaryotic. The purpose of
this report is to differentiate between the two cells by discussing the principal features of each.

Figure 3.3   Introduction to a Report

The title and the first line in Figure 3.3 clarify the subject—*Cells*. The problem that created this writing situation was a college assignment to write a paper on *Cells;* the assignment demanded the differentiation between the procaryotic and the eucaryotic. The subject of a paper and the problem that created the need to write are therefore closely allied.

Purpose

Many authors reveal their purpose in writing a paper by saying, "The purpose of this paper is. . . ." (See Figure 3.3.) This phrase is useful to readers and writers alike. The readers are happy because you have satisfied their curiosity about what is going on. As the author, you can feel confident you have included key information in the introduction.

A purpose statement does not have to be so obvious, however. Analyze Figure 3.4.

Rocks can be classified into three types based upon where they are formed: igneous—rocks formed in the earth; sedimentary—rocks formed in the water; and metamorphic—igneous or sedimentary rock exposed again to heat and pressure

Figure 3.4   Example: Introduction

The author might have written, "The purpose of this paper is to classify rocks on the basis of where they were formed"; however, the purpose—to classify rocks—is easily recognized in the present version without this type of direct commentary. Either method is acceptable although the second is more economical.

The introduction found in Figure 3.4 may have originated from any of several situations. It could have resulted from a student's effort to solve a classroom problem—writing a classification paper. It could have resulted from a geologist's problem—explaining an elementary classification of rocks to a beginning geology student.

Scope

Readers also want to know the boundaries or limitations placed on the discussion. This limitation is the *scope*.

Scope is established by defining and qualifying until the readers know what to expect. The introduction found in Figure 3.3 on cells establishes the scope as two types: procaryotic and eucaryotic. Figure 3.4 establishes the scope by classifying rocks into three categories: igneous, sedimentary, and metamorphic. In each case, the reader understands how broad or narrow the scope of the discussion will be. Figure 3.5 contains an additional example.

With all the dishwashers on the market, buying a good, reliable model at a reasonable price is a challenge. This paper evaluates three models—ScrubKing SK 1000, Acme A95, WDC C95—on the basis of price, washing performance, and maintenance record.

Figure 3.5   Example: Introduction

Rather than discuss the over eighty models and brands available, the author of this introduction limited the scope to three machines, an adequate sample for a short document and a limited survey. The reader, of course, benefits from this information (scope) and has a realistic expectation about what will appear in the resulting document. The author's problem, the one that *informs* this paper, apparently concerns selecting a good dishwasher.

Organization

Each of the sample introductions discussed so far in this chapter also suggests the organizational pattern of the paper that was ultimately written. The statement of organization (sometimes called a *road map*) is a listing of the main points discussed. Of course, the scope statement often does much the same thing, making it difficult to tell where one ends and the other begins. This really does not matter; what *is* important is giving clear signals to the reader about what to expect—in what order—in the body of the paper.

Here are some typical examples of statements of organization:

Example A: This report will be divided into four major areas: 1) —————, 2) —————, 3) —————, and 4) —————.

and

Example B: The most important aspects of the subject are —————, —————, and —————.

Although the four elements of an introduction can occur in any order, it is particularly helpful to conclude an introduction with the statement of organization. In this way you can clearly signal the structure of the body.

This textbook uses the four-part introduction as its standard pattern of organization. Check the first page of this or any other chapter for further examples. Because this is a sound, useful introductory pattern, it is also appropriate for all writing assignments in this book. In fact, it is the specified pattern for each major waiting exercise.

DIVIDE OR CLASSIFY YOUR SUBJECT AND GIVE CLEAR SIGNALS —————————

The statement of organization discussed in the last section is based upon the premise that you can divide or group ideas in order to discuss them. Instead of dealing with the entire subject, you write about a series of smaller units. In fact, most business and technical subjects lend themselves to division or classification. The following list suggests a few of the possibilities:

- recommendations and proposals
- findings, conclusions, questions
- processes, techniques, instructions
- trends, changes
- summary, background, history
- arguments (pro/con)
- alternative programs, methods, and approaches

In a practical sense, then, you take the overall subject and divide or group it into several parts. Next you list them in the order in which you plan to discuss them; this becomes the statement of organization in the introduction. A subject that is divided into parts is much easier to handle.

The statement of organization in Figure 3.6 signals what is to come and at the same time provides key words that are useful as transitional signals to organize the body of the paper. (Key words have been capitalized to emphasize them for this example.)

Although you may not want to use such an obvious link to the statement of organization in everything you write, consider this approach as fundamental to signaling organization. Use it frequently.

ARTICULATORY SPEECH PROBLEMS

Of all the speech problems the classroom teacher encounters, the most frequent are articulatory, when speech sounds are not produced in an accepted manner. One way to classify articulation disorders is on the basis of the type of error. The most common are omissions, substitutions, and distortions [This sentence is the statement of organization. It groups disorders into three categories.]

[The body of the report begins here. The first sentences of subsequent sections are quoted to show their relationship to the statement of organization.]

If speakers omit sounds normally present in pronunciation, they have an articulation problem characterized by omissions. . . .

In substitutions speakers substitute one sound for another or substitute an incorrect sound for a normal one. . . .

Distortions occur when the correct sound is approximated but not closely enough to be normally accepted. . . .

Figure 3.6   Introduction: Relationship to Body of Paper

Organization of the body of a paper is directly based on what is said in the introduction. If you are able to write a clear four-part introduction, you will probably have little difficulty writing the body of the paper.

Subsequent chapters in this book include details about how to organize the introduction and body for specific writing assignments.

CONSIDER USING A
SUMMARY BEGINNING

The four-part introduction is a good standard form; however, many variations are possible. One of the most common, the summary beginning, is simple to use, helpful to the reader, and emphatic.

Some busy company officials prefer the executive summary (Figure 3.7), a variation of the summary beginning. It gives key information quickly, allowing the reader to decide whether to read the rest of the document.

I. Executive Summary

This report summarizes project activities from May 17 to June 20, 1986.

As of June 20, work on the project was estimated to be 43 percent complete compared to 37 percent scheduled. The estimated labor budget used through the same period was $139,596 or 43 percent of the total approved budget.

The remaining work to be performed under Work Order 1431 will be completed as scheduled by September 1, 1986.

Figure 3.7

It is easy to image how useful such an introductory summary is to a busy executive facing four or five reports, ranging from ten to one hundred pages. The executive summary in Figure 3.7 prefaced a list of tasks; this list constituted the body of the report, and was followed by a short conclusion.

Although an executive summary will often pull key data from the body of a report to emphasize it, as an introduction to the report the summary must also clarify at least the subject and purpose (see the first sentence in Figure 3.7). This is also true of the three variations discussed in the following sections, each of which can be a basic form of a summary introduction:

1. a brief statement of the main ideas
2. a digest of all sections of the paper, or
3. an abridgement of results, conclusions, or recommendations.

A Brief Statement of the Main Ideas

Many writers consider a brief statement of the main ideas to be the most helpful beginning. It works well with nearly all plans of organization, and in its most abbreviated form is very similar to the thesis idea discussed in Chapter 2. It is helpful to the nonexpert and the specialist alike because they can read a memo or report intelligently and rapidly.

Figure 3.8 shows a negative example (A) of a chronologically arranged trip report that buries important information in the middle. It is followed by a positive example (B) that stresses the key findings.

```
Version A: On August 1, 1986, Tom G., Vernal M., and Joe F.,
and I met with John H. and Mary S. of the Denver OSM staff. . . .

Version B: The Denver OSM staff indicate they will approve
our new mining plan as soon as we file a response to the Complete-
ness Review. This report examines specific points discussed
in the meeting and suggests a tentative timetable for submit-
ting the Completeness Review.
```

Figure 3.8

The first sentence of Version B states the main idea derived from the meeting. The second sentence clarifies most of the standard introductory elements: subject and purpose in writing the report, and, to a lesser extent, the scope and organization.

Digest of All Sections of a Paper

A digest or listing of all sections of a document—sometimes called a descriptive summary—is particularly helpful if you need to discuss or group unrelated subjects. As Figure 3.9 shows, a digest is really just a detailed statement of organization functioning as an introduction.

Although you may not understand the subject matter discussed in Figure 3.9, you have little doubt about what will follow: one section on *economic factors*,

Subject: Minimum Reserves for 80-Acre Well Justification.

The first part of this report addresses the economic factors that remained constant throughout all of the CW35 runs—initial investments, dates, prices, down-time, and minimum flows. Next, the report addresses items involved in a sensitivity analysis—operating costs, decline rates, initial flow rates and reserves. The last section explains the attached figures.

Figure 3.9   Digest Introduction

another on *sensitivity analysis*, and a final section explaining *attached figures*. Indirectly, you can also recognize that these three points are the subject of the paper, that the author's purpose is to discuss them, and that the scope of the paper is limited to these points.

An Abridgement of Results, Conclusions, Recommendations

In papers recommending action or citing results and conclusions, a brief summary or abridgement (see Figure 3.10) in the introduction provides readers with a helpful frame of reference for reading the remainder of the document. The executive summary cited in Figure 3.7 is a variation of this form.

Evaluation of CW-35 shows the well is capable of producing 250 BOPD. This study recommends installing a holding tank and using trucks to transport the crude to the terminal. This study reviews the history of CW-35, evaluates three possible development plans, and concludes with a recommendation.

Figure 3.10

The primary difference between a "A Brief Statement of the Main Ideas" and the present pattern is content. Instead of reporting the key idea behind a study, "An Abridgement . . . ." stresses the results, conclusions, and recommendations of a study. Sometimes the two patterns overlap to the extent that it is impossible to distinguish between them.

CONCLUSION

In your college classes or on the job, use the four-part format—subject, purpose, scope, statement or organization—as the basic pattern for reports and other documents. After you have mastered its use, you will probably want to add some variation to your writing, depending on circumstance. For example, if the four-part format is too heavy for a short memo, shorten it to just subject and purpose. On other occasions, you may have a key recommendation that you will want to mention at the beginning. Many other variations are possible. Remember, your goal is to let readers know immediately what is going on and then to keep them informed about where they are in the discussion.

Most writing assignments in this book specify a four-part introduction or minor variation as the standard format.

REVIEW QUESTIONS

1. What do readers want to know as they begin reading?
2. What are the four basic statements made in an introduction?
3. Is it acceptable technical writing style to use this statement: "The purpose of this paper is to. . . ."? Explain.
4. Why is it important to clarify the *scope* of a paper?
5. What is a *statement of organization?*
6. How does the statement of organization help clarify the organization of a paper?
7. What is the typical form of a statement of organization?
8. How are division and classification useful approaches to organizing a paper?
9. What is a *summary beginning?*
10. What are the advantages of the summary beginning?
11. What are three common forms of the summary beginning?
12. What is the particular benefit of the *digest* approach?
13. What is the relationship of this chapter to the writing assignments included in this book?

EXERCISES

1. Read the following short introduction and identify the four basic elements of an introduction:

   Acme Oil facilities have experienced thousands of dollars of loss due to inadequate control of the equipment yard. This paper is a study of several alternative methods to reduce the loss, including increased company surveillance, employment of a private security firm, and/or construction of a fence and check point. This report recommends construction of the fence and check point. Following consideration of the three alternatives are a cost comparison and summary.

2. Identify the following points, if they apply, in the introduction to a fifteen-page engineering report:

    1. Subject
    *2. Purpose of research
    *3. Purpose in writing the report
    4. Scope
    5. Plan of development
    6. Background and perspective

    * Points 2 and 3 are important. *Purpose* may refer to the technical investigation and /or to the reason for writing the document. Often an introduction will discuss both points.

INTRODUCTION

As a result of moderately severe flooding on the XXX and XXXX Rivers, damage was inflicted to several spur dikes in the XXX area and at a number of locations along the XXXX River.

The design purpose of these fields of spur dikes is to maintain high-velocity flow channels away from the pipeline right-of-way and thus prevent erosion of pipeline cover and support material. The dikes are designed to resist erosive attack and remain intact even in the event of severe flooding.

The fact that these spurs did not perform satisfactorily in their first operational season under the attack of only moderate flooding, raised questions concerning the viability of the spur concept to protect the pipeline. To deal with these questions, a task force was established to determine:

1) The specific causes of the spur failures on the XXX and XXXX Rivers.
2) The impact of these failures on the viability of the spur field concept throughout the project.
3) Requirements for design modifications that would rectify problems identified in 1) and 2).

This report presents the findings of the task force. It is based on the results of a field inspection carried out by selected members of the task force, on data from surveys carried out in the spur area, and on various discussions that have taken place over the past several weeks.

3. What problem within the company caused the author to write the document cited in question number 2? What was probably the audience? What evidence do you have for your opinions?

4. Analyze the following introduction to a research report entitled "Which common Personal Computer Is Best for a Land Surveyor?" Identify the same points requested in question 1.

INTRODUCTION

During the past decade computers have greatly changed the way we work in our everyday jobs. One rapidly expanding area is the use of computers to assist land surveyors in their work. With all the computers on the market today, many surveyors are confused about which computer will best fit their needs.

The purpose of this paper is to learn which common personal computer best fits the needs of a general land surveyor in private practice.

Following a short introduction on what surveying is and how computers assist surveyors in their work, this paper evaluates four personal desk-top computers against each other in the following areas: central processing unit, progammability, internal memory, program storage, keyboard, graphics, and price. The four computers compared are the _____, _____, _____, and _____.

Each chapter includes a short introduction followed by a comparison and recommendation. The paper concludes with a short summary and final recommendation.

5. What problem in the company caused the author to write this document? What is the intended audience? What evidence do you have for your opinions?

6. Your supervisor has asked you to write an inspection manual for a pipeline suspension bridge on a major river crossing. Because failure of the bridge could cause a major oil spill and great environmental damage, it is essential to develop a regular inspection program and carry it out.

   After much consideration, you have decided to include sections that describe a Yearly and Five Year Inspection Plan. You also realize it will be necessary to provide a general description of the structure, including General Layout Drawing No. 73–157–1. One of your main items of graphic support is a proposed survey layout that can be used with the yearly inspection for discovering any movement in the foundations.

   Directions: Write a four-part introduction based upon the data presented here. It may not be necessary to use all the data.

7. For additional practice and discussion regarding the writing problem and audience, analyze the introductions in Figures 3.8, 3.9, and 3.10.

8. Turn to the introduction of this chapter and analyze it for the four basic parts of an introduction.

4

Visual Support

As you complete the assignments in this text, in other college classes, and later on the job, much of your success in communicating ideas clearly may depend upon your use of effective visual support. Charts, maps, photographs, graphs, tables—anything that helps the reader *see* is invaluable in written communication.

Great economy and clarity can be achieved with an effective combination of written language and a visual representation. For example, if you were writing instructions on how to solder a wire to a lug, just defining the terms and distinguishing between a good and poor solder joint would take several paragraphs of careful explanation. With all the written explanation, many readers would still have a problem "seeing" what you are talking about. Yet, Figure 4.1 and one or two sentences of explanation can accomplish the job quickly and effectively.

This chapter is an introduction to visual support for technical writing. The purpose is to help you visually support your writings. Discussed are Audience Analysis; Basic Theory and Standards; Tables; Graphs; Drawings and Diagrams; Visuals—Making Your Own; Photographs; and Computer Generated Visuals.

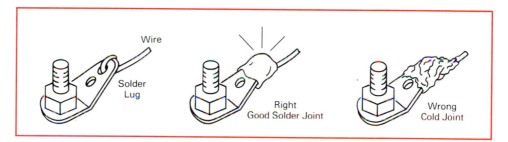

Figure 4.1   A Good Solder Joint

AUDIENCE ANALYSIS

The need to analyze audience is the same for both writing and visuals. The central question to ask youself is *What will enable me to communicate best with my audience?* Choose visual support that provides insight to the written text, clarifies the relationship of parts, and makes the trends or relationships of data instantly clear.

Audience analysis also directly influences how you present data. For a technical audience, statistical data presented in a table is attractive because it is detailed and accurate. A nontechnical or semitechnical audience however, may be turned off by all the precise data, but may be pleased with the rounded-off values presented in an easy-to-read bar chart.

BASIC THEORY AND STANDARDS

This section explains *white space*, tells how to tie visuals to the written text, and lists guidelines for including visuals in your papers.

White Space

Graphics specialists like to talk about *white space*—those parts of the page that are not covered with words or visuals. Have you ever started to read a single-spaced typewritten page with narrow margins and no indentations—just a sea of type? Your first reaction was probably, Do I *have* to read it? The same document, with attractive margins, double-spacing with paragraph indentations, a few single-spaced lists—in other words, some *white space*—is positively inviting by comparison.

Relationship: Visuals to Text

All visuals, whether included in the body of a paper or in an appendix, must be referred to in the written text. As a general rule, place them as close as possible to where you mention them in the written text, but never before they are mentioned (this way you avoid confusing the reader). Use specific explanatory sentences—i.e., "Figure 4 shows this trend for 1980–88"—to lead the reader to the appropriate visual.

If a visual contains complex information, explain it in captions or within the text. However, avoid the pitfall of writing every detail in the text and then repeating yourself in detail in the visual. For example, if a graph shows a trend over a ten-year period, in the text you should draw the reader's attention to key data. However, do not repeat every single piece of information. You want to inform your readers, not bore them.

Guidelines for Visuals

Good visuals—like good writing—are consistent in format, terminology, and style. The guidelines listed here are generally consistent with those used by the author and by student authors in model papers appearing in this text. Use them for all visuals you include in your writing:

1. Label tables as Table 1, 2, 3, and so on, consecutively throughout the text. This label is centered *above* the table or is flush left (even with the left margin).
2. Label all other visuals—charts, graphs, drawings, diagrams, illustrations, photographs, maps, etc.—as Figure 1, 2, 3 consecutively throughout the text. This label is centered *under* or flush left *under* the visual.

3. Center or place flush left under the Figure or Chart number a descriptive title, like "Summary of Operations for 1986" or "Trends in Interest Rates."
4. Make the visual self-explanatory: identify units (metric tons, volts, dollars); label columns on tables; label axes (vertical and horizontal scales on graphs); include necessary information but avoid clutter.
5. Orient labels horizontally for easy reading.
6. Keep visuals simple by focusing on one point.
7. If data for the visual or the visual itself has been borrowed from a source—magazine, book, etc.—document it. Practice varies in different professions. This text conforms to the standards of the Modern Language Association (MLA). MLA requires the author's name (title, if an author is not listed), followed by the page number. For example, if an article was written by Sam F. Jones, the documentation would be the following: From Jones (208).
8. Place documentation below the visual or table. Leave a double space below the figure number and title.
9. For contrast and emphasis, place the visual, in a box. For typed papers, use a pen and ruler to enclose the visual. If you type on a computer, use something like an asterisk to create the box, or add the lines later with a pen and ruler. Place labels, headings and documentation immediately below the box.
10. Because your papers and visuals should be as professional-looking as you can make them, do not use pencil in the final product.

Figure 4.2 is a line graph that follows these guidelines.

TABLES

Tables, lists of data usually arranged in rows and columns, are useful for showing large amounts of specific, related data in a brief space. Tables are more concise than written text and more accurate than graphs because so many facts can be conveyed at once. The arrangement of data in rows and columns (Figure 4.3) makes comparing figures easy and, of course, creates white space and variety.

Tables are not effective for showing overall trends; graphs and charts do a much better job. Nontechnical audiences sometimes find tables difficult to read; these readers prefer the more easily read chart or graph.

Guidelines for Using Tables
1. Begin each column with a heading to identify data. Study the headings in Figure 4.3.
2. Use standard symbols and abbreviations to save space.
3. Use decimals instead of fractions, unless common use dictates otherwise.
4. If you must divide a table to continue it to another page, repeat the column headings.

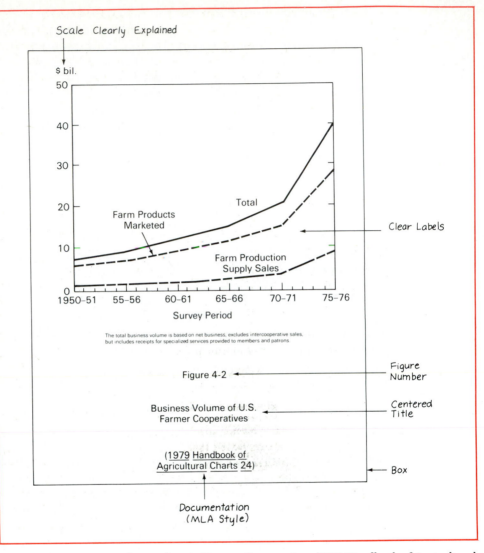

Figure 4.2   Business Volume of U. S. Farmer Cooperatives (1979 Handbook of Agricultural Charts 24)

5. Include all information necessary. For example, if you are citing data on sources of electric generation, and do not include data on hydroelectric plants, your table will not be useful.
6. If a table is located lengthwise on the page, place the top against the inside binding.
7. Follow the general guidelines for visuals concerning labeling, titles, documentation, and clarity.

Table 1

Window Energy Demand by Orientation (KBTU/Sq. Ft.)

CITY	NORTH		EAST		SOUTH		WEST	
	Wntr	Sumr	Wntr	Sumr	Wntr	Sumr	Wntr	Sumr
Dallas								
No Film	− 24	−95	+26	−156	+102	−118	+26	−185
Low-emissivity	− 7	−80	+35	−135	+107	−101	+38	−161
Reflective	− 46	−41	−33	− 56	− 14	− 46	−33	− 63
New York								
No Film	− 84	−43	−38	− 76	+ 29	− 59	−38	− 81
Low-emissivity	− 43	−39	− 2	− 68	+ 58	− 53	− 2	− 73
Reflective	−105	−11	−93	− 19	− 76	− 15	−93	− 20

Figure 4.3   Example: Table (Courtesy Department of Commerce)

GRAPHS

A *graph*, broadly defined as a means to present numerical data visually, may include line, bar, and pie graphs, as well as many others. Graphs have several advantages over tables and data: they more clearly indicate movements, distributions, and trends while presenting data in an interesting, easily recognizable form. However, graphs are less accurate than tables. Therefore, where great accuracy and visual impact are required, both graphs and tables should be used.

Line Graphs

You probably studied graphs in your mathematics and science courses. They are based upon a horizontal line (X axis) and a vertical line (Y axis). The horizontal line or scale usually represents time, and the vertical represents amounts or numbers of something (people, temperature, loans, deaths, etc).

Distortion of data is likely if the scale on both axes is not chosen carefully. By altering either scale the plotted line moves up or down. Study Figure 4.4.

The data represented by A takes a middle ground; B, with its condensed horizontal scale makes the change seem dramatic; C with a condensed vertical scale makes the change seem dramatic; C with a condensed vertical scale makes the change seem minimal. Your job is to select the scale that most fairly represents your data. Some unethical writers, seeking to mislead readers, distort a scale deliberately.

Guidelines for Preparing Line Graphs

1. Follow the general guidelines for labels, titles, documentation, and general format.

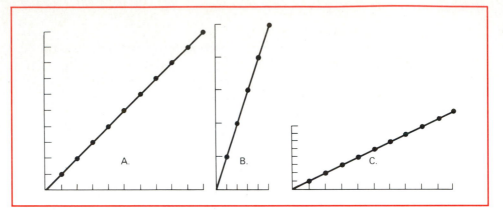

Figure 4.4   Impact of Scale Changes

2. Place the independent variable, usually *time*, on the horizontal line; the dependent variable should be placed on the vertical line. See Figure 4.5.
3. Label both variables clearly.
4. Choose scales that accurately represent the trend in the data.

Figure 4.5
Monthly Ground Surface
Temperatures (Courtesy
Department of Commerce)

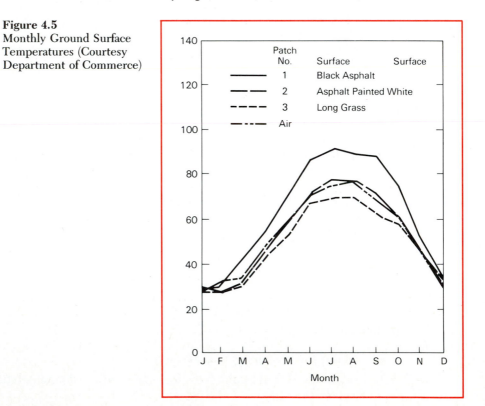

5. If you plot more than one line, limit the number to three or four and label each clearly. Study Figure 4.5.

Bar Graphs

Bar graphs are horizontal or vertical bars that vary in length to represent a quantity. Vertical bars (Figure 4.6), sometimes called *columns*, are useful to compare magnitude, size, or to emphasize difference in an item over time.

Figure 4.6, a bar or column graph, shows the difference in world exports of one item over an eight-year period. The vertical and horizontal scales are clearly labeled and the meaning of each column is also clearly explained, making the graph self-explanatory.

Figure 4.7, a horizontal bar graph, shows magnitude, size, or emphasizes difference, particularly among many items, at one specified time.

Figure 4.8, a variation of the bar graph format called a 100 percent bar or column graph, compares parts of an item that make up the whole. The bar is equivalent to 100 percent and the parts or sections represent portions of the item. This graph is easily read and has high visual impact.

Guidelines for Constructing Bar Graphs

1. When comparing an independent series of data over time, keep the graph simple by limiting the number of bars in one group to three.

Figure 4.6
World Cotton Exports (Courtesy Department of Agriculture)

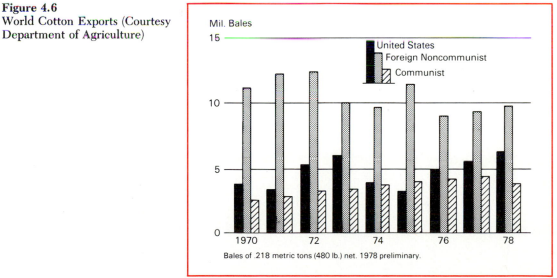

Figure 4.7
Share of After-Tax Income
Spent on Food

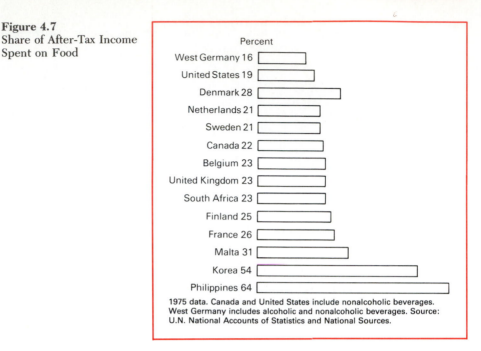

1975 data. Canada and United States include nonalcoholic beverages.
West Germany includes alcoholic and nonalcoholic beverages. Source:
U.N. National Accounts of Statistics and National Sources.

2. If you need to portray positive and negative values—often called a deviation bar graph—place negative values below the 0 reference line and label clearly. See Figure 4.9.
3. Follow general guidelines for visuals concerning labeling, titles, documentation, and clarity.

Figure 4.8
Water Source for Western Crop Acreage
(Courtesy Department of Agriculture)

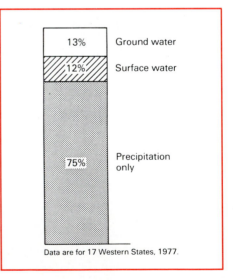

Data are for 17 Western States, 1977.

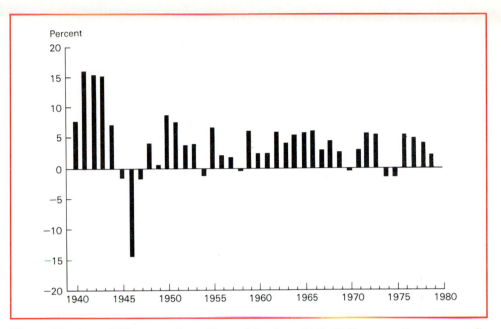

Figure 4.9   Annual Change in Gross National Product, 1940–79 (Courtesy Department of Agriculture)

Pie Graphs

A pie graph is a version of the 100 percent bar graph. The percentages, however, are represented as wedge-shaped sections of a circle. The circle equals 100 percent of a single item—dollars, hours spent, and so on. Each piece of the pie shows how that single item is divided. Figure 4.10 shows which portions of household energy consumption in 1976 were devoted to various uses. Although the sections are difficult to compare, the pie graph is dramatic and easy to read because of its strong pictorial quality.

Guidelines for Constructing Pie Graphs

1. Start the largest portion of the pie at the "twelve o'clock" position and move clockwise, going on to increasingly smaller portions. Every 3.6 degrees equals 1 percent.
2. Avoid small portions (.5 percent, etc.) that won't show up well. Put the small portions together in a "miscellaneous" or "other" portion.
3. Follow the general guideline for visuals concerning labeling, titles, documentation, and clarity.

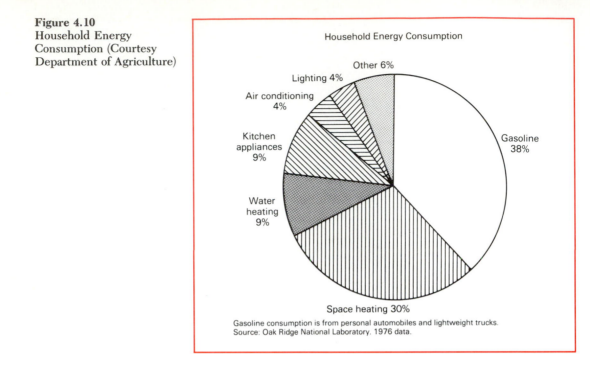

Household Energy Consumption

Other 6%

Lighting 4%

Air conditioning
4%

Kitchen
appliances
9%

Water
heating
9%

Gasoline
38%

Space heating 30%

Gasoline consumption is from personal automobiles and lightweight trucks.
Source: Oak Ridge National Laboratory. 1976 data.

DRAWINGS AND DIAGRAMS

A drawing usually portrays the actual appearance of an object. Unlike a photograph, a drawing gives you complete control over what you include. This flexibility
allows you to match the visual perfectly with the content of the text. For a
reader who must understand the physical characteristics of a mechanism to
use it properly, a good drawing is valuable.

Your ability to control the content of a drawing gives you extreme flexibility
in the form of cutaway, exploded, and cross-sectional views.

Cutaway Drawing

A cutaway drawing removes a portion of a mechanism's exterior, revealing the
relationship of inner parts to one another and to the exterior housing. Figure
4.11 suggests the advantages.

Exploded Drawing

An exploded drawing, as the name implies, breaks a mechanism into parts but
maintains an orderly arrangement, showing the parts in appropriate sequences.
An exploded drawing (Figure 4.12) is particularly useful in explaining how to
assemble or repair a mechanism.

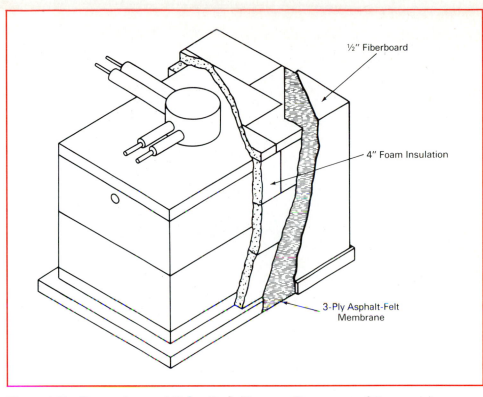

Figure 4.11   Precast Sectional Utility Vault (Courtesy Department of Commerce)

Cross-Sectional Drawing

A cross-sectional drawing (Figure 4.13) shows a mechanism cut in half, revealing both the exterior and interior. A cross-sectional drawing and a cutaway drawing differ mainly in the size of the portion removed.

Diagrams

A diagram is actually a drawing—a plan or sketch—consisting mainly of lines and symbols. It is used to explain how something works, what it looks like, and where the parts are located in relation to the whole. Diagrams are widely used in engineering, manufacturing, and construction. Figure 4.14 is a simple example.

A specialized variation is the schematic diagram common to technical fields such as plumbing, electrical wiring, and electronics. It differs from other diagrams

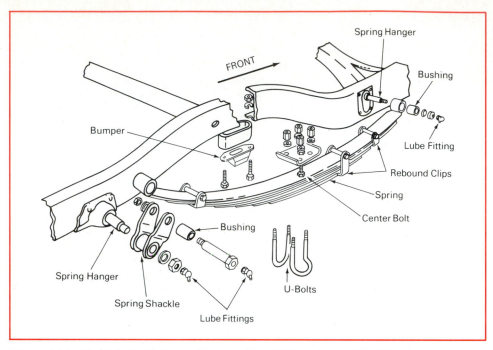

Figure 4.12   Multiple Leaf Spring (Courtesy Naval Education and Training Support Command)

Figure 4.13
Simple Carburetor with
Throttle Valve (Courtesy Naval
Education and Training and
Support Command)

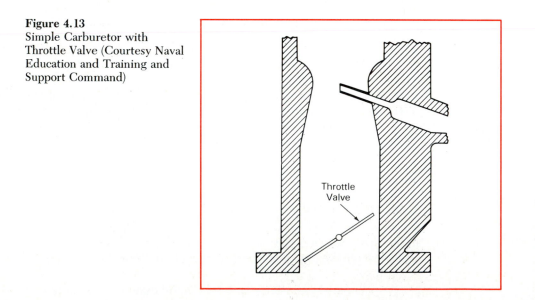

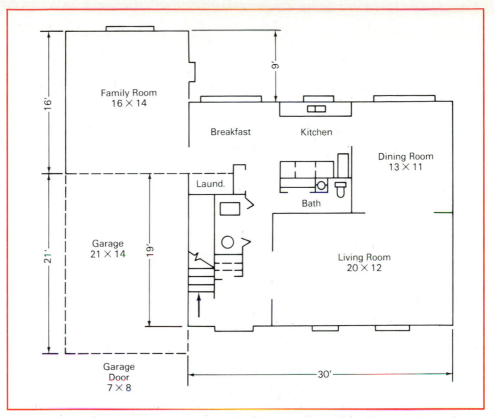

Figure 4.14 Simple Diagram of the 1st Floor of a Family Dwelling (Courtesy Department of Commerce)

because it does not attempt to portray literal appearance. Figure 4.15, for example, uses symbols, lines, and arrows to show how a dual liquid solar hot water heater works, but has little or no similarity to actual hardware.

Other commonly used schematic diagrams are organizational and flow diagrams (charts). See Figures 4.16 and 4.17, respectively.

Guidelines for Making Drawings and Diagrams

1. Make drawings simple.
2. Be consistent in maintaining proportions and size.
3. For complex drawings, provide appropriate keys and explanations. Use symbols and explanations your audience will understand.
4. Follow the general guidelines for visuals.

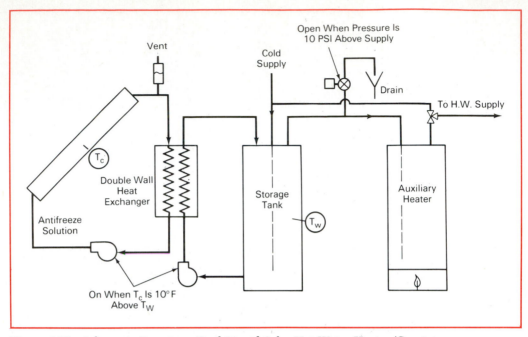

Figure 4.15   Schematic Drawing—Dual Liquid Solar Hot Water Heater (Courtesy Department of Commerce)

VISUALS—MAKING YOUR OWN

Although creating some illustrations may require considerable experience, you can provide satisfactory visual support for your writing using simple drawing tools and aids, commonly available pens, and a photocopy machine.

Simple drawing tools and aids, easily purchased in businesses that supply drafting supplies, will help you avoid the wobbly lines of free-hand drawings and give visuals a professional touch. For example, a *t-square and right triangle* are useful for aligning visuals with the margins of a page and for drawing lines parallel or at right angles to one another. For drawing a variety of different sized circles, a simple *circle compass* is handy. For curved lines that the circle compass cannot produce, a *French curve* allows you to make neat, firm lines. Many symbols such as those used in electronic schematics, computer flow charts, and home-layout diagrams are easy to make using *templates*, pieces of plastic with cutouts. To use one, you simply place a pen point in the cutout and mark around the edges. Figure 4.18 will familiarize you with these items.

Rub-on letters, also sold with drafting supplies, are easy to use (see Figure 4.19). After selecting type style and size, you transfer the letters to your visuals

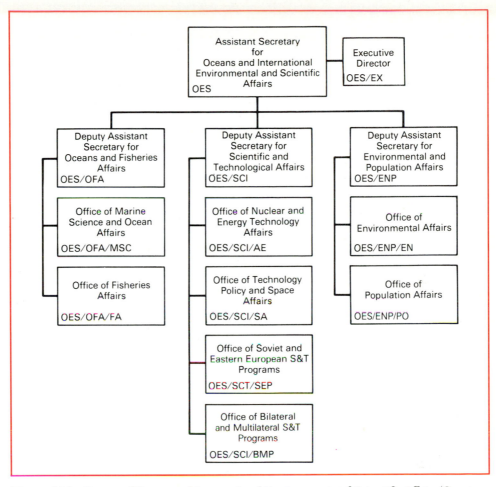

Figure 4.16 Bureau of Oceans and International Environment and Scientific Affairs (Courtesy Commission on Organization of Government for Conduct of Foreign Policy)

by rubbing them with a soft pencil. If you use a light pencil line for alignment, the result will be professional-looking captions and labels.

Although technical pens for drafting are readily available, many ordinary pens now available will work nearly as well. For most line drawings, choose a pen with black ink, capable of making fine or very fine lines. For colored visuals try some pens of the soft-tipped variety.

Perhaps the most versatile tools for producing visuals are a pair of scissors, some transparent tape, and a photocopy machine. While preparing a document you may find an illustration in a book or periodical that fits your purposes precisely or that will fit if modified properly. A photocopy machine is the key to success.

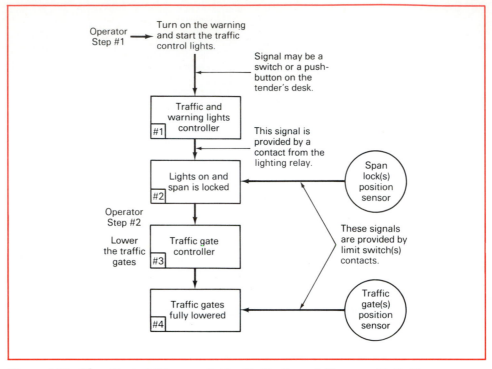

Figure 4.17 Flow Control Diagram, Bridge Traffic Control (Courtesy U. S. Department of Transportation)

Here is the basic process:

1. Photocopy the visual.
2. If the image is too large or too small, use the copy machine to reduce or enlarge it. (Some machines will vary the copy size from 64 percent to 141 percent of the original (see Figure 4.20).
3. Write down all data needed to give credit to your source. Often items may require written permission from the publisher! Because many materials published by the federal government are usually not covered by copyright laws, they are particularly good sources.
4. Trim the visual to fit your purposes.
5. On a separate sheet of paper, draw a box to outline the visual. Type in the appropriate figure number and title, or use rub-on letters.
6. Place the visual in the box and select a position to display it effectively.
7. Use transparent tape on all sides of the visual (this will eliminate unwanted lines on the final copy).
8. Photocopy the visual and new background to produce the finished copy (Figure 4.21 shows a visual prepared by this method).

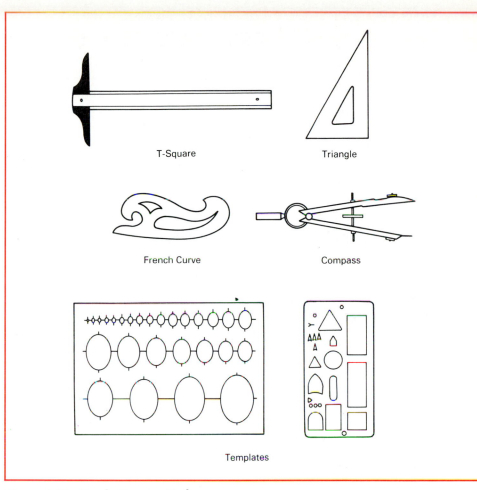

T-Square

Triangle

French Curve

Compass

Templates

Figure 4.18   Simple Drawing Tools

Figure 4.19
Samples: Rub-on Letters

	Size	Type Style
A B C D E F G H I J K L M	12	Garamond
A B C D E F G H I J	14	Schoolbook Italic
A B C D E F G H I J	18	Kartoon
A B C D E F G	20	Megaron Bold
A B C D	24	Eurostile Bold Extended

Figure 4.20
Photocopy: Reducing/
Enlarging

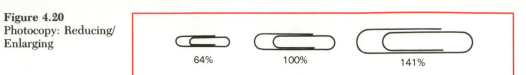

64%   100%   141%

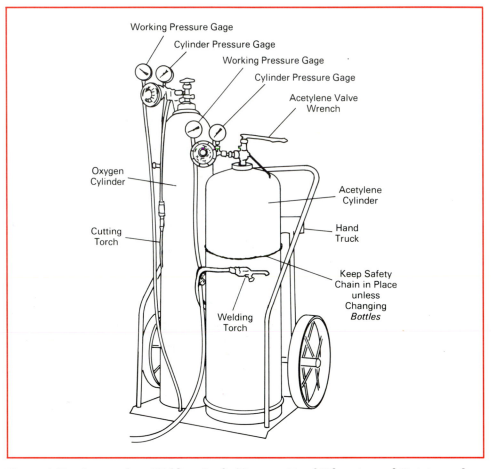

Figure 4.21   Oxyacetylene Welding Outfit (Courtesy Naval Education and Training and Support Command)

You can also make "photo" copies by placing an object directly on the glass of the copy machine and lowering the lid. By experimenting with the contrast control, you can quickly produce a surprisingly good "picture" as shown in Figure 4.22. Of course, once you have the image on paper you can use the photocopy machine to enlarge or reduce it to fit your text.

Figure 4.22
Visual: Prepared by Photocopying an Object

PHOTOGRAPHS

Like drawings, photographs are useful to show how an object looks. If you mention *oxyacetylene welding outfit*, some readers will not have the slightest idea of what you mean. A photograph (Figure 4.23) will quickly make the term

Figure 4.23
Photograph: A Good Method to Show What Something Is

Figure 4.24
Distance Perspective: **a.** Too Close **b.** About Right
c. Too Far Away

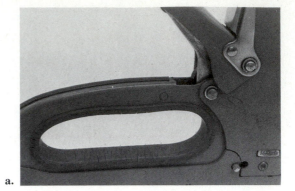

a.

b.

c.

specific and concrete. If you are doing research on water pollution, a photograph of dead fish floating on the water with a known landmark in the background will provide strong visual evidence to support your text.

Although a photograph is ideal for some situations, it can have limitations. Its strength—showing what something is—can also be a weakness. The viewer may see many distracting details that are not directly related to a specific subject. For example, a photograph of the instrument panel in a small airplane shows everything at once, even though the text is dealing with just two instruments, the altimeter and turn and bank indicator. A drawing, by contrast could feature an outline of the panel and highlight only the instruments being discussed.

Guidelines for Photographs

1. Get as close as you can. Most amateur photographers get too far away for good detail (see Figure 4.24).
2. To clarify size, place a ruler, coin, or other known object next to the item being photographed (see Figure 4.25).
3. Avoid distracting backgrounds by controlling what is included in the photograph and/or by cropping. Simply cut away what is not helpful.

Figure 4.25
Ruler Indicates the Relative
Size of Object in Photograph

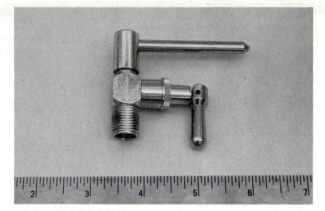

4. For a three-dimensional perspective, take the photograph at an angle that catches some top, end, and side views (see Figure 4.26).

COMPUTER-GENERATED VISUALS

Graphs and other computer-generated visuals are increasingly common in papers written by college students. You may already own a micro-computer and/or have access to a college computer laboratory that has graphics programs capable

Figure 4.26
Camera Angle Creates a Three-
Dimensional Effect

Figure 4.27
A Student Generates a Pie Graph
on a Computer

of producing all illustrations in this chapter. Hand-drawn visuals, while still common in student papers, are rapidly being replaced in the classroom and in business by computer-generated visuals.

Not only can computers produce excellent visuals, but they also do it quickly and accurately in black and white and stunning six-color combinations. Just a few years ago an executive wanting a colored graph portraying company earnings in a given year had to contract the graphics department and wait days, even weeks, for delivery. Because of the lead time involved, labor costs, and general difficulty in obtaining visual support, many executives, engineers, technical writers, in the past used little graphic support. Those days are over: executive and college student alike can generate a colored graph in just minutes with a microcomputer.

Many experts feel that the graphics revolution is just beginning, and that far more flexible combinations of visual materials will follow. This is an exciting prospect but one that could lead to abuse. Remember, when choosing visual support the central question to ask is, *What will communicate best with my audience?* Too many visuals or inappropriate visuals have the power to hinder communication. Seek a proper balance of written text supported by appropriate visuals.

If you have the computer equipment and software to generate visual support for your college courses, use them to support your writing. If you do not have this access, drawn them by hand. Whatever the case, effective technical writing requires visual support.

REVIEW QUESTIONS

1. What is the central question to ask when choosing visual support for written text?
2. What is *white space?* What is its value in visual support and in written text?
3. What is the proper relationship between written text and visual support?
4. How should you show the specific relationship between the written text and a figure?
5. Where should the *Figure* label appear in a visual?
6. Where should you place the documentation in a figure?
7. Why is documentation required for some visuals?
8. What does the text recommend as a good method to gain contrast and emphasis for a visual and its labels, headings, and documentation?
9. What are the strengths and weaknesses of tables?
10. What is the main advantage of graphs over tables?
11. How can an improperly drawn vertical or horizontal scale distort the accuracy of a line graph?
12. Bar graphs are particularly useful for portraying what kind of information?
13. What is a 100 percent bar graph? How is it similar to a pie graph?
14. What feature particularly distinguishes a *deviation* bar graph from other bar graphs?
15. What are the chief strengths and weaknesses of photographs as visual support?
16. How can you clarify for a viewer the relative size of an object in a photograph?

EXERCISES

General Directions: Use the general Guidelines for Visuals (see p. 48) to prepare the following exercises. Use the specific guidelines for each particular type of visual. All answers should include appropriate headings, labels, and figure references.

1. Find a convenient location to monitor motorized (or foot traffic) past your residence. Group your observations in some convenient pattern such as the number of cars, vans, pickups, trucks, semi-trailers, motorcycles—and the overall total number of vehicles. In a table, present your data and what percent of the total each category represents.
2. Make a line graph showing the following data (use the horizontal scale for time). Jack W. Wand earned the following yearly grade-point averages (four-point scale) over six quarters of school.

- Fall 1985   2.5
- Winter 1985   2.75
- Spring 1985   2.89
- Fall 1986   2.29
- Winter 1986   2.9
- Spring 1986   3.2

3. Construct a bar graph comparing tuition increases in three private colleges over a ten-year period. Use the following data (data equals cost per semester):

COLLEGE A	COLLEGE B	COLLEGE C
• 850.00	760.00	1000.00—1974
• 850.00	860.00	1200.00—1976
• 925.00	960.00	1350.00—1978
• 1125.00	1300.00	1350.00—1980
• 1125.00	1225.00	1500.00—1982
• 1200.00	1225.00	1600.00—1984

4. Estimate your expenses for a week, month or semester and present the information as a 100 percent bar graph.

5. Present the data used in Problem 4 as a pie graph.

6. Construct a cutaway drawing of a simple mechanism related to your major field. An electronics student, for example, could draw a potentiometer revealing the slide and resister element normally concealed within the housing.

7. Construct a cross-sectional drawing of a common mechanism such as a spring-operated ballpoint pen.

8. Draw a floor plan for your home or housing unit. Include details such as doors, windows, bathroom, and so on.

9. Draw a flow chart for the registration process used at your college. Assume a novice audience who are not acquainted with the process.

10. Draw an organization chart for a college club or for the administration of your college.

11. Make a visual for one of your writing assignments, using a photocopy machine. Reduce the visual and provide your own labels and documentation.

12. Using photographs from your own collection, prepare a brief document showing photographs taken too close to a subject, at an appropriate distance, and too far away. Support the photographs and write a brief analysis.

PART II
WRITING BASIC PATTERNS

5

Definition

Definition, clarifying the meaning of specialized terminology, is an important part of technical communication. In an age of high technology, the need for definition arises with each new advancement and each new generation of readers. Those making new discoveries and creating new processes must communicate them to management, to the general public, and to people involved in the technology of other fields.

For example, a specialist in designing cells i.e., (the micron-level parts of fingernail-sized electronic chips that are used in computers and electronic games) is thoroughly familiar with technical terms and should use them when writing to other specialists. The choice of words in Figure 5.1 suggest a specialist audience:

```
We assume that you have experience in logic and circuit design
using standard TTL or CMOS logic families, and will extend
this experience to allow you to use NMOS Standard Cells.
'' . . .'' Input and Output (I/O) cells allow you to easily
interface with TTL or CMOS circuits, and microprocessor tri-
state data bases. Worst case gate delays as low as 10nS per
level, and flip-flop toggle frequencies of 16MHz are provided.
```

Figure 5.1

However, to an administrator with a business background, to a reader in another discipline, or to the company executive making decisions about buying new equipment, this passage means nothing. To communicate with these varied audiences, the electronics engineer must select material carefully. Terms *must* be defined.

This chapter discusses informal, formal, and extended definition. Examples written by students show how to use these forms and/or combinations of them in writing a definition paper.

INFORMAL DEFINITIONS

Informal definitions include the *synonym*—using a familiar word for an unfamiliar one; an *operational definition*—telling what something does; and an *illustration*—using a drawing or photograph to clarify meaning. Although less precise than formal definitions, informal definitions are more common because they are easy to write and because they fit into the flow of ideas smoothly.

Synonym

A synonym is a word that means the same thing as another. By substituting a familiar word for the one you think the reader may not know, you often avoid the need for additional definition. For example, instead of *numismatist* you might say the more familiar *coin collector;* instead of *hypertension* you could write *high blood pressure;* instead of *matrix,* use *chart* or *list.*

Operational Definition

An operational definition states what something does. For example, a *CRT* gives a visual representation of a computer program; a *sphygmomanometer* measures blood pressure; and a *numismatist* collects coins.

Illustration

Perhaps one of the simplest ways to define a part or clarify a concept is to include an illustration. By examining a drawing, particularly an appropriately labeled exploded view, a reader can grasp a term quickly. For example, the simple drawing of a bicycle in Figure 5.2 clarifies the relationship of parts and the meaning of such terms as *stem bolt, head tube,* and so on.

Informal definitions—synonyms, operational definitions, and illustrations—are appropriate for many situations because they occupy little space and do little to disrupt the flow of ideas.

Figure 5.2
Defining through Illustration

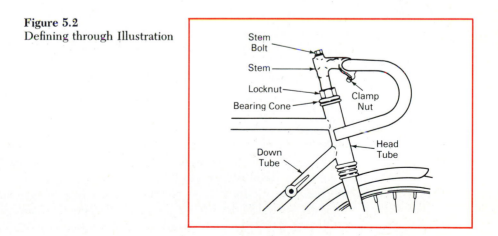

FORMAL DEFINITIONS

Formal definitions offer precision. They are written in a traditional form containing an item, class, and differentiation. The *item* names what you are going to define; the *class* puts it into a particular group; and the *differentiation* separates it from all others of its group. Figure 5.3 contains several examples artificially separated to show the three parts:

```
A computer ──[item]──→ is a machine ──[class]──→ which receives,
stores, manipulates, and communicates information
    ──[differentiation]──→ .

A CPU ──[item]──→ is a microprocessor ──[class]──→ which performs
basic arithmetic logic in the entire system
    ──[differentiation]──→ .
```

Figure 5.3   Item—Class—Differentiation

Formal sentence definitions sometimes appear, as you may have noticed, in a characteristic sentence pattern or frame. To write your own definition, substitute words appropriate to your subject in the blanks (see Figure 5.4).

```
A _____ is _____ which/that _____ .
```

Figure 5.4   Sentence Frame

A formal definition does not have to be a single sentence; it can be as long as necessary to define a term clearly for a given audience. What makes it a formal definition is the attention paid to clarifying the item, the class, and the differentiation.

The following example (Figure 5.5) from *Webster's Collegiate Dictionary*, contains the item, class, and differentiation. It is a good example of the types of formal sentence definitions that often appear in technical documents.

Notice how the differentiation characterizes a typical instrument. To save space the verb is replaced with a colon. As is often the case, the *which/that* are also deleted. After the sixth word the author might have said *which/that*

Thermometer: an instrument for determining temperature consisting typically of a glass bulb attached to a fine tube of glass with a numbered scale and containing a liquid (as mercury or colored alcohol) that is sealed in and rises and falls with changes of temperature.

Figure 5.5 Dictionary Definition

consists, but chose not to. Either the standard form characterized by the sentence frame or a slightly modified version such as in Figure 5.5 is widely accepted.

GUIDELINES FOR
WRITING FORMAL DEFINITIONS

Here are a few guidelines for writing polished formal definitions:

1. Make the class as specific as possible. For example, if you say a "thermometer is a device that" . . . you have not accomplished your purpose, because many things are devices. You have not narrowed the class at all. It is better to say "a temperature-sensing device" or better yet "a temperature-measuring instrument that. . . ."

2. Use the same part of speech for the *class* word as you use for the item. For example, if the item is a pen, use a noun such as *writing instrument* when you state the class.

3. Do not say "A thermometer *is what/occurs when.* . . ." Such expressions sometimes appear in informal definitions but have no place in formal definitions.

4. Make the differentiation specific enough to separate the item being defined from all or most others of its class. For an example, see the thermometer definition in Figure 5.5.

5. Avoid circular definitions that repeat the word being defined in the item and then again in the class or differentiation. This is unacceptable form in most cases. For example, do not say that a "thermometer is an instrument that measures the thermo quality of a liquid or gas." The reader still needs a definition of *thermo.*

6. If the differentiation cannot be completed in one sentence, add what is necessary. If you add very much, the result is called an *extended* or *amplified* definition.

WRITING AN
EXTENDED DEFINITION

An *extended* or *expanded* definition, as the name suggests, involves more than one sentence. Depending upon the writing situation, audience, and the need for detail, you may write a paragraph or even several pages to clarify a definition. After reading a formal sentence definition, an uninformed audience often needs to know more. Extending may be done with one of several methods, and frequently a combination of methods helps to clarify the point. Some of the most common methods are derivation, explication, exemplification, analogy, comparison–contrast, and analysis.

Derivation refers to the origin of a word. For example *Webster's Collegiate Dictionary* lists the derivation of the term *thermometer* as from the Greek *therme,* meaning "heat," and from *meter,* meaning "measure." In a similar sense, an acronym can often be clarified by telling what it means: *LOFT* means Loss Of Fluids Test.

Explication means to explain any words in the definition that may confuse the reader. If, for a novice audience, you defined a microscope as "an optical instrument that consists of a lens or combination of lenses for making enlarged images of minute objects," you would follow with an explanation (explication) of the definition. *Optical instrument* may need further clarification; *lens* may need further definition, and so on. Discuss each key word in the definition until all have been clarified. If needed, clarify further with an example or other information.

Exemplification, or giving an example, is one of the best methods to extend a definition. After defining *merit system* as "a method of recruitment that uses an examination to discover which persons are most qualified for a particular job," you might follow with an example such as this: "When a job opening comes up, interested persons report to the Office of Personnel Management to take the examination. The manager in charge reviews the top five scores and after an interview with each, weighs their qualifications and decides whom to hire."

Analogy is a comparison of the unfamiliar with the familiar, based upon a pattern of similarities. Analogy often helps the reader relate to the term being defined. One writer, referring to a dressmaker's *seam gauge* used this simple analogy:

The seam gauge resembles a short ruler, with a lengthwise slot in the middle running almost the entire length.

A *Comparison–Contrast* definition shows similarities and differences by focusing on something the reader knows in order to explain something else. The emphasis may be on similarities, differences, or both. Figure 5.6 emphasizes contrast:

A personal computer is a small computer based on a micropro-
cessor; it is a micro computer. Not all microcomputers, however, are personal computers. A microcomputer may perform a
single task such as controlling the temperature in a building;
it can be a word processor, a video game or a hand-held four-
function calculator. A personal computer is a stand-alone
computer that makes a wide range of capabilities available
to an individual. A personal computer is a system that has
the following characteristics: The price of a complete system
is less than $5,000. The system typically includes disks as
a form of secondary memory.

Figure 5.6   Definition by Contrast

Figure 5.6 initially emphasizes contrast between a personal computer and other forms of microcomputers. Because the author wanted the term *personal computer* clearly defined, he or she specifies two specific characteristics in this passage and further develops the contrast in subsequent paragraphs.

An extended definition may also be developed by analysis into parts and particular models or versions, by cause-and-effect analysis, by illustration, or by any combination of methods that will clarify the meaning for the reader.

STUDENT MODELS—
EXTENDED DEFINITION

The writing situation for each of the following three models was the same: the students were required to write an extended definition (approximately 150 words) for a novice audience.

Each model uses three-part organization. The formal sentence definition constitutes the thesis idea and introduction. The supporting information is the body. The last sentences, in Models 1 and 3 particularly, serve as the conclusion, a brief signal telling the reader that "this is the end of the document."

Model 1 begins with a formal sentence definition; it then develops the body with description, analogy, and application; and concludes with a generalized comment that functions as a clincher sentence.

MODEL 1   **POTENTIOMETERS**

A potentiometer, more commonly called a "pot," is an electromechanical device containing a resistance element that is contacted by a movable slider.

Potentiometers are designed so their resistance values can be changed easily with a manual or automatic adjustment. Their most common shape is a round cylinder similar to six quarters stacked on each other with a steel shaft protruding from the center (see Figure 1).

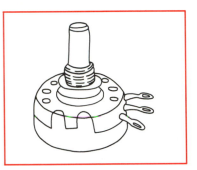

Figure 1   Potentiometer

Three terminals, located on the outside of the cylinder, are for attaching wires. The heart of a potentiometer is its resistance element, a form wound with resistance wire.

A potentiometer can be compared to a water faucet. When a faucet is wide open, water is allowed to flow with little restriction, but as the faucet is turned off, it restricts the water flow. A potentiometer works on a similar principle only it restricts current flow.

Potentiometers are used in radio and stereo equipment to control the volume, in blenders to control speed, and in dimmer switches to control light intensity in a room. Due to their adjustability, they have become an important component in electrical circuits.

Model 2 begins with a formal sentence definition and then extends the definition by analysis and exemplification.

MODEL 2   **DEFINITION OF STYLE NUMBERS**

A style number is a set of digits attached to clothing by a manufacturer or retailer to indicate variations in material, design, cut, and color. Attached to the clothing either by the manufacturer or the retailer, the number can be broken down into sets of digits which specify the different aspects.

The material or cloth is represented by one or two digits at the far left of the number; there is usually little or no variation of material in a given style of a brand. Design is represented by one or two digits to the right of the material digits, and variations of cut follow in the next two. Color, the last digit at the far right, specifies a characteristic with the greatest variation.

The style number for a pair of Levi brand straight-leg, adult-cut jeans is 5190917. The five represents denim material, nineteen specifies straight-leg design, 0917 identifies the pants as adult-cut. The entire number is printed on a leather patch above the right back pocket.

Style numbers are attached to the clothing in specific places. In pants it is either above the pocket, or in the inside lining; in shirts it is on the inside of the collar, or on the end of either the shirt-tail or the sleeve.

After a formal sentence definition, Model 3 explains the source or derivation of the word, compares it to a bank safe, and analyzes its construction and method of operation. The last sentence is a good "clincher," providing a clear signal to the reader that this is the end of the definition.

MODEL 3   COOLGARDIE SAFE

A Coolgardie Safe is a cooling device used in the desert areas of Australia to refrigerate food where there is no ice or electricity. Its primary usage was in the late eighteen to nineteen hundreds, though it is still used in some "outback" areas today.

Coolgardie is the name of a gold-rush town about five hundred miles from the coast of Western Australia. In 1890 miners flocked to the "golden" town and the problem of cooling food became a major concern. A box, resembling a bank safe, using water and air, was invented to cool the food. Thus, the name "Coolgardie Safe" was given to the invention.

The "safe" is constructed of wood, burlap, rags, a pan, and a hose of trickling water. (See Figure 1.) A box frame is made from the wood and the burlap is stretched over all six sides. This is the part resembling the bank safe. One of the sides is slit down the middle so food may be placed inside. The pan is located on top of the safe and the rags are draped out of the pan onto the burlap. The hose is placed into the pan and the water is trickled into it. The pan eventually fills with water, soaking the rags, which in turn soak the burlap. The best location for the safe is under a large tree or someplace where it can be shaded. The contents of the safe are cooled as breezes pass through the wet burlap.

As power lines eventually reached towns like Coolgardie, the safe was replaced by the electric refrigerator. In small "outback" towns and sheep-camps,

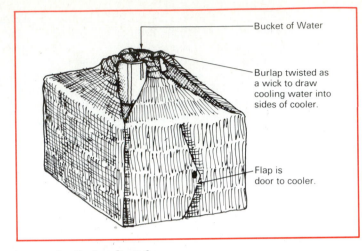

Figure 1   Coolgardie Safe

the safe is still in use. Though primitive and almost extinct, the Coolgardie Safe was and still is an effective cooling device.

REVIEW QUESTIONS

1. When is definition needed in writing?
2. Why is an informal definition often preferred over a formal definition?
3. How can a synonym be used as an informal definition? Give an example.
4. What is an operational definition? Give an example.
5. How can a picture be used as an informal definition?
6. What are the three basic parts of a formal definition?
7. What is the simple sentence *frame* that suggests the basic form of a formal sentence definition?
8. What are the six guidelines for writing formal definitions?
9. What is an extended definition? How does it differ from a formal sentence definition?
10. What are some of the common methods of developing or extending a definition?

EXERCISES

1. **Writing Formal Definitions.** Write formal sentence definitions for the following terms:
   a. screw driver
   b. floppy disk
   c. mixer (stereo equipment)
   d. sprocket
   e. cumulus cloud
   f. allergy
   g. "cool" (slang term)
   h. retarded
   i. power-to-weight ratio
   j. magnetism

2. Choose five terms common to your major field and write a formal sentence definition of each. Assume a college-level audience.

3. Assume an audience of sixth-grade students. Show how you would adjust your approach and style to this audience by writing brief definitions (one–five sentences) of the five terms used in Exercise 2.

4. Using the six guidelines discussed in this chapter, evaluate the following formal definitions. Make any corrections needed to produce an adequate formal definition.
   a. A temperature inversion occurs when the temperature aloft is higher than on the ground.
   b. A technical pen is a drawing pen used by draftspersons.
   c. A milling machine is a machine tool.
   d. Osteoporosis is a disease of the bone.
   e. A table saw is a sawing tool used for sawing wood and other materials.
   f. A computer is a device.
   g. A sod stripper is for stripping sod.
   h. An error message is where the operator has done something wrong.
   i. A "stroke" is a swing of a golf club.
   j. A "hookshot" is a shot where the arm is positioned like a hook.

5. **Writing an Extended Definition.** Write an extended definition (100–250 words) of a concept or item for an intelligent reader who is not acquainted with the subject. For a writing situation, assume you have been assigned by your employer to write a series of documents to explain key company terminology to new employees.

   Use a formal sentence definition in the first paragraph (introduction). It may be necessary to add an additional sentence or two of explanation to orient the reader properly. See the first paragraph of Models 2 and 3 for examples.

   The body of the paper, probably two–four paragraphs, will include different methods of development: explication, exemplification, analogy, comparison/contrast, and so on. Think of your readers' needs. They will want information such as "What does it look like?" "About how big is it?" "What is it for?"

A short paper such as this will have a minimal conclusion, probably no more than a single sentence that moves the focus of the definition away from the specific back to the general. The last sentences in Models 1 and 3 are good examples.

An appropriate subject for this paper could be a term associated with your major field. If you are interested in computers, try *ROM* or *RAM* or *modem* or *bus control*. If you are majoring in automotive technology, *dwell*, or *camshaft*, or *float* or hundreds of other terms are available. Every field has its jargon, terms that only insiders know.

6

Description of a Mechanism

This is the second of the chapters dealing with specific writing techniques used in technical communication. Writing a description of a mechanism is a common activity for technical writers, especially as parts of longer reports. *Mechanism*, broadly defined here as almost anything that has a specific purpose, could range from a jumbo jet to a simple paper clip.

Anyone who must share information about equipment, programs, or how to perform a specific task will need to describe a mechanism. Figure 6.1 is an excerpt from a letter written to a federal regulatory agency answering an inquiry concerning new pipeline inspection equipment:

```
In answer to your question regarding new hardware, we are re-
ferring to the new corrosion deformation inspection tool oth-
erwise know as the C-D Pig*. The C-D pig utilizes a one-piece
non-articulated frame. The pig is propelled through the pipe-
line by the oil flow and is rigidly supported on the pipe center-
line by four urethane cups. A bumper at the nose protects the
pig from the in-line check valve clappers. The main body is
a pressure tight housing for the power supply, the tape re-
corder, and the electronics.
```

Figure 6.1   Description of a Mechanism. (*A "pig" in pipeline jargon is an inspection device inserted into a pipeline.)

This chapter explains how to write the three basic parts of a description of a mechanism: the introduction, part-by-part description, and conclusion. Also included is a discussion of analogies, an outline showing typical organization, two model papers, review questions, and writing assignments.

WRITING THE INTRODUCTION

Meet the needs of readers by using the standard four-part introduction combined with an illustration and/or initial visual image.

State the *subject* and *purpose* to explain the "what" and the "why" of your paper. To give readers a clear initial understanding of the subject include a *picture* and/or *description* of the object in broad detail. To indicate the *scope* and *plan of organization*, devide the subject into a list of parts that you will discuss in the body of the paper. The next paragraphs discuss each of these points in greater detail.

To introduce the *subject* and *purpose*, you may choose a direct approach with a simple statement: "The purpose of this paper is to describe a common plumbing fitting, a ¾″ cast-iron pipe cap." A slightly less wordy version might say, "This paper describes a common plumbing fitting, a ¾″ cast-iron pipe cap." The "purpose of this paper is . . ." statement, as discussed in Chapter 3, is particularly useful in writing a first draft because it forces you to include key information in the first paragraph, right where your readers need it.

The *initial visual image*, which helps readers "see" the object, can be a photograph, drawing, and/or word picture (see Figure 6.2). The illustration should be simple, placed close to the introduction, and might include a known object (such as a ruler) to help readers understand relative size.

If a photograph or drawing is not available, help readers "see" the mechanism by specifically discussing characteristics such as shape, color, and size. Analogy, comparing the unfamiliar to the familiar, is a particularly valuable way to inform a reader. For example, ¾″ cast-iron pipe cap could be described as follows:

> This handy plumbing fitting when placed in the palm of the hand with the opening facing upward appears to be a small, heavily built metal bowl about one inch in diameter.

This simple statement helps readers' "see" by telling the relative size (fits in palm of hand) and orientation (facing upward); by explaining how the object compares to a familiar object (small, heavily built bowl); and by giving a specific measurement (1″ in diameter).

Complete the introduction by telling readers what to expect in the paragraphs and pages, to come. Breaking the mechanism into a number of parts and listing them clarifies the *scope* and *plan of organization* of your discussion. Breaking and mechanism into a number of parts also simplifies your task as a writer: instead of trying to write about the entire mechanism at once, you discuss only one part or section at a time.

Breaking a mechanism into parts is sometimes an arbitrary task, especially

Figure 6.2   ¾″ Cast Iron Pipe Cap

if the mechanism is all one unit that lacks easily recognizable parts. This really need not be a problem, however. Most mechanisms have sides, a top, and a bottom—at least four basic parts. As the author, you can choose how to divide the mechanism. But be sure your divisions do not confuse the reader.

Figure 6.3 contains each of the elements discussed in this section.

```
    The purpose of this paper is to describe an unknown object.
When placed in the palm of a hand with one of the open ends
facing downward and the other facing to the left or right,
the object resembles a small gray metal periscope 1¾" high
and ¾" in diameter.
    For this description the object consists of four main
parts: the exterior portion between the openings, the end
facing horizontally like the lens of a periscope, the inte-
rior, and the end facing downward.
```

Figure 6.3   Sample Introduction

The content of the introduction and your general approach to the subject depend upon analysis of audience. For most writing assignments included in this book, assume your readers are intelligent but uninformed about your topic. However, in some writing situations you may have an expert audience. For example, if you were writing a description of the same ¾" pipe cap for an experienced plumber, you could assume the reader has complete knowledge of the fitting. It is likely that you would not use this topic at all or, if you did, you would not emphasize physical description and analogy.

USING ANALOGIES

Audience analysis is a key factor in choosing and using analogies in technical communication. When you write for an audience that has little or no knowledge of your subject, an analogy can be an effective means to help your readers "see" it. For an expert audience, however, a simple analogy may seem trite or out of place.

Some technical writers consider analogies to be too imaginative, a poor substitute for the precise, concrete description they value in technical description. Others say that writing must be suited to your audience, and that you should use whatever works. A good approach is to use analogies selectively and to avoid highly imaginative comparisons.

When you choose an analogy for your writing, try to be as factual and concrete as possible. In Figure 6.3, a ¾″ cast-iron pipe elbow (the unfamiliar) is compared to a miniature periscope (the familiar). This analogy works because the shape of a periscope is well known to most readers. The pipe elbow might also have been compared to the bowl of a smoker's pipe with the stem missing; in fact this version would likely have been more effective because it expresses similarites in both shape and physical size.

If you have an object that does not lend itself to an effective analogy, or requires a highly imaginative comparison, create your initial visual impression by including specific measurements and details.

WRITING THE
PART-BY-PART DESCRIPTION

The part-by-part description constitutes the body of the document. As always, you must have the readers' needs in mind. If you have divided an object, a hammer for example, into two parts—head and handle—what do readers need to know? In a sense it is like starting over, because readers need to know what the handle is, what its purpose is, and what it looks like. You may even need to do some further subdividing of parts. This need is partially countered by the knowledge readers gained from reading the introduction and seeing the drawing (or written description). Write just enough to clarify the subject and the relationship of parts.

One good way to introduce a *part* is to extend the initial analogy you mentioned in the introduction. Using the plumbing fitting (Figure 6.2) as an example, you might begin a part-by-part description by talking about the interior of the "bowl" or the bottom of the "bowl." Either version gives readers a clear idea of your subject. They have a visual reference and can keep in step with your description.

The description of the part, once you have introduced it by tying it to the initial analogy, should feature careful description of such aspects as the following:

- color
- shape
- size
- material
- relationship to other parts
- texture

How exacting should you be in your description? Audience analysis and the purpose of the description give the best clues. For a college writing class, your instructor will probably expect measurements correct to 1/16", supported by considerable detail.

In addition to careful description and measurement, orientation is also important. Orientation is the position of the mechanism in relation to the viewer. In the general introduction you should, if possible, have given the mechanism's position as you began the description. The 3/4" cap, for example, was "in the palm of the hand with the opening facing upward." This, of course, is a necessary position for the "bowl" analogy to be meaningful.

As the part-by-part description develops in the body of paper, you may want to turn the mechanism on its side to examine the interior; be sure to tell readers about the change! For example, for an interior view, you might write: "With the object lying on its side with the opening facing the viewer, the flat surface of the bottom is clearly visible." Keeping readers informed about orientation is an important part of a description of a mechanism.

A final consideration in writing the body is the *person* or *point of view* to use. Should you use *I* to refer to yourself in the text, or should you avoid personal pronouns (I, you, he, she, it, we, you, they) altogether? This, of course, depends upon audience. For formal technical papers, as a general rule avoid the first person *I*. For more informal documents such as letters and memos, the first person is usually expected and accepted. Most major writing assignments in this book are formal. When you are writing on the job, ask your supervisor or another employee what company policy is, and follow it.

WRITING THE CONCLUSION

If a description of a mechanism is part of a longer work, a conclusion often is not needed. However, if the description is a separate document, as it is in the writing assignments at the end of this chapter, you have an obligation to wrap up the discussion.

To choose an appropriate ending, consider the readers' needs once more. Have you clarified how the mechanism works or what its purpose is? Is a short summary helpful? In some papers this is important data. Often the simplest (and best) ending is to state in a direct manner that you have reached the end. You could include a statement such as "This concludes the description of. . . ."

Study the writing models at the end of this chapter, to see how the authors concluded their papers.

Assignments in this chapter focus on static description. Chapter 8, on process description, explains how to describe a mechanism in operation.

COMMON FORMAT

The following outline (see Figure 6.4) demonstrates a good way to organize the description of mechanism papers you may be asked to write.

```
1.0 Introduction
    1.1 Subject (Name the mechanism if you recognize it)
    1.2 Purpose (To describe an object)
    1.3 Initial visual representation
        1.3.1 Approximate size
        1.3.2 Initial orientation of the object
        1.3.3 Analogy to something familiar
    1.4 Part-by-part division

2.0 The Body (Part-by-part description)
    2.0 Introduce the part
        2.1.1 Repeat the name
        2.1.2 Clarify the orientation if it has changed
        2.1.3 Tie the part to the initial analogy
        2.1.4 Divide into subparts if needed
    2.2 Describe it carefully
        2.2.1 -etc. size, shape, color-
    2.3 Introduce part 2, etc.

3.0 Conclusion (May include some of the following)
    3.1 Purpose of the object
    3.2 Use
    3.3 Summary of main parts
    3.4 Description of the object in operation
    3.5 Conclusion statement-"This ends a description
        of . . ."
```

Figure 6.4   Sample Outline—Description of a Mechanism

MODELS OF STUDENT WRITING

This section presents two different approaches to a writing assignment in which students were asked to write a description of a mechanism. The first, a description of a known object, features visual support; Figure 1 provides the initial visual image. This paper follows the outline discussed in the last section. The conclusion is a simple generalization that pulls the reader away from the detail of the body.

In the second paper, the author had to assume that the mechanism is unknown. This, of course, eliminates any visual support and forces the writer to concentrate on careful description, analogy, and basic writing skjills. Because the object is unknown, the writer also has to avoid discussing use and purpose. The second paper follows the sample outline and concludes with a simple statement of completion.

Although somewhat unusual in business or technical writing on the job, both versions provide valuable practice in careful observation and accurate writing.

MODEL 1: KNOWN OBJECT

THE "TOT" STAPLER

I. Introduction

The "Tot" stapler is a small mechanical device used for small fastening jobs. Staples (small pieces of precision bent wire) are ejected by the stapler, through the paper, and crimped on the other side of the paper to make fast, easy joints.

The overall device is only slightly larger than a regular-sized clothes pin: three inches long by one inch high. It has three major parts: the main body, a plastic cover, and a base plate hinged to the back of the body (see Figure 1). These parts work together to produce either crimped or straight staple joints.

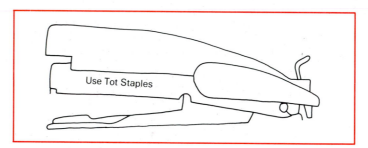

Figure 1   "Tot" Stapler

II. Part-by-Part Description

Body.   The main body of the stapler is hollow, open from the top for staple loading. This cavity is approximately one-half inch wide and two inches long, just wide enough for the row of staples to fit inside. Towards the back of the cavity are a pressure rod and spring, used to keep the row of staples pushed snugly into place at the front of the cavity (see Figure 2.).

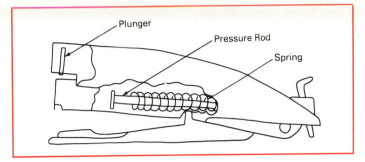

Figure 2   Cutaway View

A release button, connected to the back of the body and projected through the plastic cover above, is conveniently located for easy use. To open the body cavity, push the release button, which releases the tension of the pressure rod, allowing the plastic cover to slide back revealing the inner cavity. To close the mechanism, simply push the cover back in place; the release button automatically snaps back into place.

Plastic Cover.   The top piece of the stapler is smooth hard plastic used as a lid to protect the body. The cover slides away to reveal the inner cavity when the release button is pressed.

Toward the front on the inside bottom of the cover is a small metal plunger (Figure 2). Pushing the plastic cover down depresses the plunger, ejecting one staple through a narrow slot at the bottom of the cavity and crimpling it on the other side of the paper.

Base Plate.   The base plate, the bottom piece of the stapler, is a stand and base to crimp the staples. The metal plate protrudes directly under the main body. Two small grooves in the plate appear immediately under the plunger. When a staple is ejected through the slot in the cavity, each side of the staple hits a groove in the base plate and crimps to form a tight joint.

Hinged to the backside of the main body, the base plate also makes it possible to eject uncrimped staples. The plate is easily snapped back into place by pushing the main body and base plate towards one another.

III. Brief Description of Mechanism in Operation

To staple something, place materials in the open slot between the body and the base plate. Push the plastic cover down, fairly hard, and the staple projects through the material. For other stapling jobs, such as bulletin boards, unhinge the base plate and press the body of the stapler directly to the object being stapled.

The "Tot" stampler is a simple, dependable mechanism that requires little skill to operate.

MODEL 2: UNKNOWN OBJECT

DESCRIPTION OF AN UNKNOWN OBJECT

The purpose of this paper is to describe an unknown object. When placed in the palm of a hand with one of the open ends facing downward and the other facing to the left or right, the object resembles a small gray metal periscope 1¾″ high and 1¼″ in diameter.

For this description, the object consists of four main parts: the exterior portion between the openings, the end facing horizontally like the lenses of a periscope, the interior, and the end facing downward.

With the object positioned so the ridged rim is facing to the right and the unmarked rim facing downward, the exterior portion of the "periscope" is visible. Its surface is a dull medium gray metal with specks and scratches of darker gray and rust discoloration. The rim of the lens opening protrudes 1/16″ from the body of the periscope and is ¼″ wide. At unequal ⅛″ to 1/16″ intervals along the rim are slashes or markings running lengthwise or at a slight angle. They vary from surface scrapes to ⅛″ deep. The area between the lens opening and the lower opening is a 90 degree angle. The lower rim of the lens opening is ½″ higher than the plane the base rests upon. An inverted raised "C" is visible in the center of the lower body, ⅜″ from the base.

With the object turned so the lens opening is facing away from the observer, two irregular flattened areas are visible on the lower body. One on the lower rim is ½″ long and ¾″ high. The second, positioned 1/16″ from the main body and is ¼″ high. It has many dents and scratches but lacks the slashes found on the lens rim.

With the object rotated so the upper opening is facing the observer, the "periscope lens" is visible. The opening is ¾″ in diameter encircled by a ¼″ thick rim. The rim's surface is flat except for a projection at the top and a minute indentation circling the opening 1/16″ from the outer edge. The indentation ends at the projection and continues clockwise after a ⅜″ gap. The projection is 1/16″ high and wide. Starting 1/16″ inward from the circular indentation, the rim slopes down at a 45 degree angle for approximately 1/16″.

The interior, seen by looking into the lens area, has ½″ of threads 1/16″ deep and spaced 1/16″. These threads are a dark gray. Adjacent to them, the interior surface of the body is almost entirely rust, covered with a few scratches where gray metal shows. The lower opening has approximately ½″ of rust-covered threads 1/16″ deep spaced 1/16″.

To view the end facing downward, the observer must rotate the object so the lower opening is directly facing him/her and the lens opening is pointing up. The lower opening, ¾″ in diameter, is encircled by a 3/16″ rim. The rim's surface has a few minor protrusions and two that are more noticeable located on opposite sides of the opening. The one on the right is ¼″ long, 3/16″ wide, and projects 1/16″ at its highest point. The smaller projection on the left is ⅛″ long, 1/16″ wide and 1/16″ high. The inner edge of the rim slopes down at a 45 degree angle for about 1/16″ .

This concludes a detailed technical description of an unknown object that resembles the curved top of a miniature periscope.

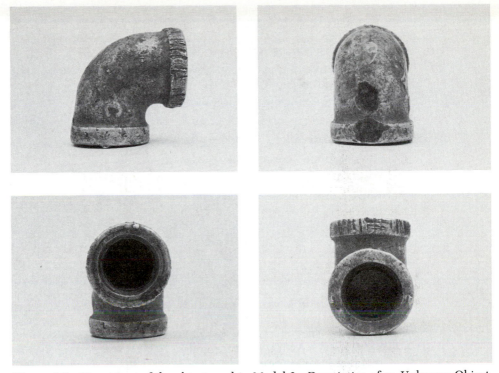

Figure 6.5   Four views of the object used in Model 2—Description of an Unknown Object

Both papers use three-part organization and clear organization. Model 2 is particularly successful in using the periscope analogy to tie the paragraphs of the body to the initial visual image.

In describing a mechanism, many students fall into a grammatical trap: they let modifiers dangle. For example, an orientation statement may read "Turning the object so the bottom is visible, a pattern of small indentations is visible." The phrase "Turning the object so the bottom is visible" is an action requiring a performer; if one isn't included in the sentence immediately following, the "Turning . . ." phrase dangles, making a mildly illogical statement. To avoid the problem, provide a subject for the phrase. Corrected, the sentence would read as follows: "Turning the object so the bottom is visible, *one* (an observer, or someone capable of "turning") can see a pattern of small indentations." For a full discussion of modifiers, see p. 268.

Figure 6.5 shows each orientation used in the part-by-part description in Model 2. Although this paper does not provide visual support, you may find it interesting to compare the pictures of the mechanism with the written text.

REVIEW QUESTIONS

1. How is *mechanism* defined in this chapter?
2. What information should you include in the introduction to a description - of-mechanism paper?
3. What is the purpose of the *initial visual image?*
4. What is the organizational basis for the body of the paper?
5. How can extending an analogy help organize the body of a paper?
6. How exacting should your measurements be?
7. What is *orientation* as it relates to a description of mechanism?
8. What is one simple way to state the conclusion?

EXERCISES

1. Analyze the following introductions taken from student papers describing an unknown mechanism. Evaluate them for standard four-part introductory elements, effective use of analogy, visual support, initial orientation, and writing style. Assume the authors were writing formal papers for an audience of college sophomores.

   a. The purpose of this paper is to describe an unknown mechanism. The mechanism looks like a small egg with two holes running perpendicular through it and grooves running lengthwise from one side of each hole to the other side. It is small enough to fit in the palm of a hand. The entire mechanism is made out of ceramic clay with a grayish-white glaze on the surface. The mechanism will be described in three parts: 1) the surface, 2) the holes, 3) the grooves.

   b. This paper describes an unknown object that looks like a small, metal version of Space Mountain at Disneyland. One could even find a resemblance to the Tinman's hat in The Wizard of Oz. The object weighs less than a pen. Probably constructed of aluminum, this silver-gray mountain measures 2 $\frac{11}{16}''$ in diameter at the base, and tapers to a height of 2½ inches, with an oval hole of $\frac{5}{16}$ at the peak. The object will be divided into four parts: base, side, top, and inside.

   c. The purpose for writing this paper is to clearly describe a specific mechanical object. The object looks similar to a small fire hydrant. I plan to give a piece-by-piece analysis by starting at the top and working my way down and then inside. The parts include the cap, base, and interior parts.

   d. The purpose of this paper is to describe an unknown mechanism. When placed in its natural orientation, it resembles a small "sundial" 1½'' high and 2½'' wide that weighs approximately 3 ounces. The paper will describe the three main parts: top, middle, and bottom.

2. Revise and edit the examples used in Exercise 1. In particular, attempt to combine sentences and avoid unnecessary repetition.

3. An analogy used in a formal technical writing paper should be conservative, more literal than imaginative. It should also function on several levels, such as size and shape and color. As an analogy, the comparison should indeed be familiar to many of the potential readers. Using these criteria, evaluate the following analogies:

   a. The mechanism under scrutiny is a grayish metal object that looks like the top portion of a periscope, the part that sticks out of the water when observations are made.

   b. This object looks similar to a miniature desk telephone receiver lying broadside to the observer with the earpiece and mouthpiece facing down.

   c. This object is similar in size and appearance to an M-80 firecracker, but with a clamp around the body and a piece of metal on the end of the fuse.

   d. This object resembles a 2″ letter *B* tipped on its right side.

   e. The rod, curved like half of an ellipse with the bottom flat and both ends rounded, has the general shape of the runner for a child's sled.

   f. When the object rests on its side, it looks much like an oversized golf tee, but is still small enough to fit in the palm of a hand.

   g. The outer appearance and size of the mechanism are much like an elongated, gray thimble.

   h. Above this plate is a cap-like projection that resembles the turret on the Union's iron-clad "Monitor" during the Civil War.

   How important was "audience" in your evaluations? Why?

4. Description is written for varying purposes. Analyze the language/word-choice of several advertisements for popular products such as food, laundry soap, automobiles, alcohol. How do the choice of words, the type of appeal, and the purpose differ from those in technical description? Submit a brief analysis of each advertisement, accompanied by a copy of the original.

5. *Major Writing Assignments.* As you read the models, you probably noticed two different approaches to the same assignment: *The "TOT" STAPLER* is a description of a known object; whereas the second model describes an unknown object. Either approach accomplishes much the same goals; the second version may force you to concentrate more on careful technical description.

   In both assignments, the choice of a topic is important. Choose something that is simple—a bit more difficult than a simple paper clip but less difficult than a stapler. Here are a few suggestions:

   • *Electronics:* toggle switch, fuse (old screw-in variety), antenna insulator, resistor, capacitor, etc.

   • *Plumbing:* nipple, cap, elbow, simple valve, coupler, collar, etc.

   • *Tools:* screwdriver, punch, nailset, hammer, socket, pliers, etc.

   • *Office:* common 2H pencil with eraser, inexpensive ballpoint pen, fountain pen, typewriter brush (for cleaning the ball or keys), bottle of correction fluid, eraser—ink or pencil—in various shapes, etc.

Figure 6.6

Some instructors have a collection of numbered items (see Figure 6.6) to loan students. The object and the paper are turned in together, making it easy for the instructor to evaluate the accuracy of description.

Whatever you choose to describe, be sure it is small and simple; this will allow you to use detail in your description.

a. *Assignment 1:* Description of a known object.   Write a 500–750-word description of a simple mechanism. Use the suggested format (see p. 91) and refer to Model 1. Use simple illustrations to support the written word. Simple line drawings are particularly appropriate. Turn in the mechanism with the paper.

b. *Assignment 2:* Description of an unknown object.   Write a 500–750-word description of a simple mechanism similar to that assigned in Assignment 1 but assume that the object is unknown. Avoid speculations about use; concentrate on careful, precise description. In the introduction simply state that the name of the object and its purpose are unknown. To help the reader "see" the mechanism, particularly emphasize the analogy by referring to it several times in your paper. Study Model 2 for additional insight.

7

Classification and Division

Classification, the orderly arrangement of related data according to a systematic basis, is fundamental to writing and learning. When you visit the library to gather information, you check the card catalogue to find a particular book. All the books are classified alphabetically on the basis of author, title, or subject. If no system of classification existed, the library shelves would be a chaos of books: one on religion, the next on weather, the next on tuning a 1955 Chevrolet, and so on.

We classify virtually everything that relates to our lives. Days, for example, are "good" or "bad" based upon the temperature, wind speed, and amount of sunshine. We classify automobiles in many different ways on a multitude of bases: number of cylinders—4-cylinder, 6-cylinder, 8-cylinder; size—micro, small, compact, full-sized; size of cylinder in cubic inches or cc's.

Division, by contrast, concerns one item and its parts. In the description of a mechanism assignment in Chapter 6, you write a description of one item. Because you cannot write about all aspects of the item at once, you divide it into parts or sections and discuss them one at a time.

Classification and division are sometimes used as the patterns of organization in entire papers. More often, however, they are used as parts of a longer work. For example, a one hundred-page environmental impact statement will probably use every method of organization discussed in this book—including classification and division.

This chapter defines basic terms, explains how classification and division are used in exposition, sets guidelines and rules for their use, cites models of student writing, provides review questions, and concludes with writing exercises.

BASIC TERMINOLOGY

Five terms—genus, species, classification, basis, and division—are fundamental to understanding the discussion in this chapter.

Genus and Species

If "office supplies" is a *genus*, then paper, pens, typewriter ribbons, paper clips, and so on are *species*. If "motorcycles" is a *genus*, then trail bike, combination, and street bike are some of the *species*. As you can see by now, a *genus* is a group or class, and a *species* is a subdivision of the group or class. In botany, for example, seven basic categories are used to identify and classify organisms: kingdom, phylum, class, order, family, genus, species. Each level is part of the broader level that precedes it, no matter the subject. The classification for a dog, for example, includes the following: Animalia, Chordata, Mammalia, Carnivora, Canidae, Canis, Canis Familaris.

Genus and species, as used in this text, require understanding of only one or two levels of relationships. Figure 7.1 shows two simple examples of genus/ species relationships pertaining to timber. Part A shows two broad categories of trees—deciduous and coniferous. Deciduous, the source of most hardwood, are broad-leaf trees that usually shed their foliage in autumn; coniferous, the source of most softwood, are needle-bearing trees that appear green all year long. Part B shows how lumber may be classified by the way a tree trunk is cut.

Classification

Classification is the grouping of things such as objects and ideas that have similar characteristics. For example, if you were asked to sort a miscellaneous collection of plumbing, electronics, electrical, and cabinet fittings such as resistors, transitors, 90 degree ¾″ pipe elbows, chrome water faucets, hinges, red plastic drawer pulls, and an electrical plug, you could classify them into piles of metal, nonmetal, plastic, and combinations of materials; or possibly classify them on the basis of color—red, brown, metallic, etc.; or classify on the basis of shape—but this quickly becomes too complicated! In short, there are many ways to group them; in this instance, classification on the basis of use or function, as the four original categories suggest, is the most practical. Classification, then, means grouping items according to some system of similarities and differences called a basis.

Figure 7.1
Genus/Species Relationships

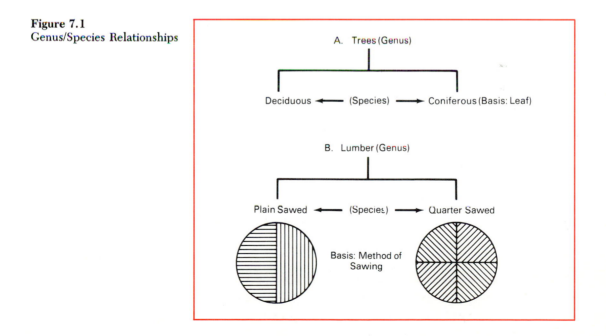

TABLE 10. Characteristics of Vegetative Habitats on the
Island Park Geothermal Area

Habitat (percent of total IPGA)	General location on IPGA	Ecological characteristics	Characteristic plant species
Douglas-fir Series (42)	• Moderate elevations (5100-7200 feet) • Broad belt which merges with mountain brush and alpine fir series	• Warmest and driest Forest areas • Lodgepole pine presently dominates most sites • Alpine fir and spruce are accidental • Aspen sometimes dominates but on small areas • Lodgepole pine is under epidemic attack by mountain pine beetle	• Douglas-fir • Lodgepole pine • Quaking aspen • Snowberry • Spirea • Globe huckleberry • Meadowrue • Aster • Sticky geranium • Pinegrass • Elk sedge
Alpine fir Series (44)	• Borders Douglas-fir Series and extends to timberline or Continental Divide • Predominantly in eastern and south-eastern portions next to Yellowstone National Park and the Teton range	• Cooler, damper sites • Lodgepole pine presently dominates and persists for longer time • Douglas-fir occurs in a seral role • Aspen, limber pine and spruce occur periodically • Lodgepole pine is under attack by pine beetle	• Subalpine fir • Lodgepole pine • Quaking aspen • Douglas-fir • Limber pine • Engelmann spruce • Globe huckleberry • Grouse whortleberry • Snowberry • Spirea • Meadowrue • Lupine • Pinegrass • Elk sedge
Sagebrush-grass (6)	• Lower elevations, on dry, coarse-textured soil • Occurs largely on Bureau of Land Management lands in central and western portions • Also found scattered throughout the forested series on dry, rocky, windswept outcrops	• Sagebrush usually dominant • Scattered Douglas-fir along southern portion along ecotone with Douglas-fir series	• Big sagebrush • Threetip sagebrush • Snowberry • Woods rose • Knotweed • Sticky geranium • Bluebunch wheatgrass • Nevada bluegrass • Idaho fescue
Mountain brush (3)	• Found on lowest and driest sites • Borders the Douglas-fir series along south-western border • Also found along river courses on dry, south-west-facing slopes	• Dominated by shrubs other than sagebrush although sagebrush is present	• Rocky Mountain maple • Chokecherry • Rabbitbrush • Big sagebrush • Serviceberry • Eriogonum • Aster • Bluebunch wheatgrass
Wet Meadow (3)	• Scattered throughout—predominantly in northern, central and southeastern portions • Found on sites with high water table, or where run-in or flooding is common	• Floristically and ecologically diverse • Highly susceptible to disturbance during growing season • Rate of recovery at higher elevations is slower than at lower sites	• Willow • Wyethia • Camass • Pondweed • Yampa • Bluegrass • Junegrass • Sedges • Rushes

Courtesy U.S. Department of the Interior

Figure 7.2   Illustration of Basic Terms (Courtesy U. S. Department of the Interior)

Basis

Basis refers to the qualities or similarities in species that qualify them to be placed in a genus or group. For a classification to make sense, the items grouped at each level must have something in common—a basis. In Figure 7.1, Part A, timber is divided into two groups—deciduous and coniferous. The *bases* of this two-part classification are primarily the leaf—broad leaf or needle—and whether the tree is green the year around. In Part B the genus *lumber* is classified into two species on the *basis* of the way the tree trunk is sawed into lumber. Documents using classification should clearly state the basis unless it is clearly implied.

Another example may be helpful to clarify the differences in meaning among *genus, species,* and *basis.* Figure 7.2 contains a table from a United States Department of Interior environmental impact statement that investigated the potential impact of drilling for geothermal resources in a heavily forested wildlife and recreational area.

The authors had complex data to present to a general audience. They pulled all their data together about *vegetative habitats* (genus) and *classified* it in five groups or *species* on the *basis* of location, ecological characteristics, and characteristic plant species. Notice how effective this table is in terms of visual impact and economy of language; it says a lot in a small space.

Division

Division is easy to distinguish from classification because of a fundamental difference between the two. In classification at least two or more items are always involved; *in division only one item is involved initially.* For example, handsaws can be classified on the basis of purpose and tooth-shape as crosscut, rip, or combination. In division, on the other hand, *one* individual saw or species (a crosscut saw for example), has two parts, a handle and blade. Study Figure 7.3.

USE IN EXPOSITION

Classification is useful in organizing data for discussion; as such, it is part of nearly every document you may write. However, you usually will not be asked to write a classification and/or division paper on the job (although you are asked to write one for practice at the end of this chapter). Classification/division is more of a technique of writing than a report form. Yet it is fundamental for you to organize your ideas into some sort of logical pattern so you can explain them clearly.

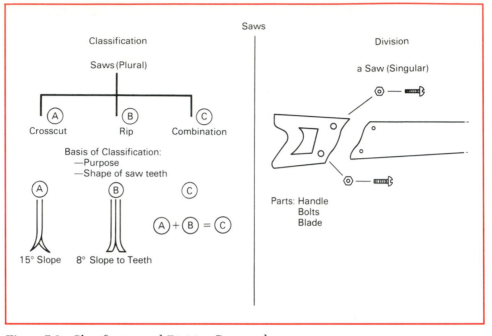

Figure 7.3   Classification and Division Compared

Figure 7.4, an introduction from a student paper, shows how classification can signal the structure of an entire paper.

The paper on cells will have two major sections or classifications—procaryotic and eucaryotic. The basis of classification will be differences in internal structure.

An amazing number of business and technical documents lend themselves to organization by classification and division. The author of a letter analyzing an off-schedule project might list thirty problems—large and small. A letter classifying or grouping the thirty observations as *labor*, *weather*, and *materials* problems would be easy to write and read. A letter attempting to list all thirty, without a pattern, would seem chaotic!

The fundamental unit of all living systems is the cell. A major
characteristic that distinguishes cells from one another is
their internal structure. On this basis, cells can be placed
into two categories: procaryotic or eucaryotic. The purpose
of this report is to differentiate between the two cells by
discussing the principal features of each.

Figure 7.4   Introduction to a Classification Paper

The front page of a daily newspaper contained an article on good places to live in the United States. After several paragraphs of highlights and background, the following sentence appeared: "The group's survey ranked cities according to 11 categories: population change, births, crowding, violent crime, individual economics, community economics, education, hazardous wastes, air, water pollution and sewage." In other words, American Cities (*genus*) are classified as good or bad places (*species*) to live based on eleven criteria (*bases*). *The Post-Register Newspaper* Idaho Falls, ID

SOME GUIDELINES

Here are some basic guidelines for writing classification and division papers. Notice how well the example on cells in Figure 7.4 relates to them.

Classification

1. Tell what is being classified.
2. State the basis of the classification.
3. Identify the classes it can be placed in.
4. In supporting discussion, use one basis at a time to maintain clarity.
5. Be sure the terms, the species and classifications, and the basis are clear to the reader.

Division

Much the same advice is appropriate for division. In addition, the following are important:

1. As you break something down into its parts, make it clear to the reader where one part begins and another ends.
2. Be consistent in handling point of view.

In division, for example, if you are writing a functional description, stay with it until you have completed the parts inventory. If you want to discuss the type of steel used in a saw blade, be sure to signal clearly to the reader that you have finished one point and are shifting to another.

MODELS OF STUDENT WRITING

The authors of the following papers wrote in response to the major writing assignment in the exercises at the end of this chapter. Both papers are successful efforts to classify data, express clear bases for the classification, and systematically discuss them in a three-part organizational format.

Model 1   **CLASSIFICATION: ARROWHEADS**

Arrowheads designed for hunting can be classified according to their shape, number of blades, and purpose. On these bases, there are three general classes: field point tip, blunt tip, and multi-blade tip.

Field Point Tip

Field point tips are generally all the same shape, having a single sharp tapered blade made of hard steel (Figure 1). Their purpose is hunting small game and target shooting. The sharp point will penetrate small animals and target bales. The hard steel will hold up against rocks and trees.

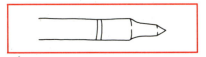

Figure 2.1

Blunt Tip

Blunt tips have two general shapes, cylinder and flared, neither of which is a really a "blade." Cylinder tips (Figure 2) are the same thickness as the arrow shaft with a flat tip. Some are solid steel and others lead-filled brass. The flared types (Figure 2) has a larger impact surface that tapers out from the arrow shaft. They are formed from either rubber or nylon. The purpose of blunt tips is to kill small game with impact rather than penetration.

A Cylinder   B Flared

Figure 2.2

Multi-Blade Tips

Multi-blade tips fall into two categories: two-blade and four-blade. Two-blade tips are available in two shapes: round point and sharp point. Round point tips (Figure 3) have a smooth razor-sharp arch shape and are made of durable tool steel. The purpose of the rounded design is to slide around bones to get to the vital areas. Sharp point (Figure 3) are diamond-shaped. Their razor edges, made from tool steel, penetrate vital organs and bones. Both round point and sharp point tips are designed to kill large game.

Four-blade tips (Figure 4) similar to the sharp pointed two-blade tips, have two additional blades on each side. The four blades are welded together

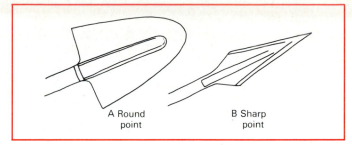

Figure 2.3

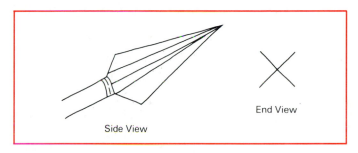

Figure 2.4

and bought together in a sharp cross-pattern point. Because of the complicated design, a softer steel is used. The purpose of the four-blade tip is to cut a large hole as it penetrates large game.

Based on shape, purpose, and number of blades, this paper has classified arrowheads into three classes: field point, blunt tip, and blade tip.

This paper has a clear beginning, middle, and end. The first paragraph clarifies what is classified (arrowheads), the basis (shape, purpose, number of blades), and the classes (field point, blunt tip, and blade tip.) The author systematically discusses the three bases in each section of the paper.

Model 2   **CLASSIFICATION: BEARINGS**

The need to overcome friction becomes greater as the technology and research in mechanics advances. One means of reducing friction is a bearing, a surface or spacer which keeps moving surfaces apart. Most bearings fit into two general categories—roller or bushing—based upon their design and purpose (Figure 1).

Figure 1   Roller Bearings

A roller bearing has two parts: the spacers and the cup/ring. The spacers are round and consist of three main shapes: rollers, balls, and needles. The needle and roller spacers, similar in shape, look like small metal rods rolling between the cup and ring as the shaft turns. A roller bearing is useful where the force of the turning shaft requires a large surface area. The needle bearing has the same purpose but is particularly useful where space is critical. The ball bearing is a round sphere that, like the rollers, rolls between the cup and ring. The ball bearing is useful where the design requires little bearing surface.

The cup and ring provide two services: they provide the surface on which the spacers travel and they retain and keep the spacers in place. The cup and rings are both circles on which the spacers ride. These rings are attached to the shaft to keep the spacers in place. This puts the force applied to the bearing at a constant direction and place. Because the spacers travel on the ring and cup, the ring and cup take up the wear and stress. This helps reduce shaft wear.

The bushing, in contrast, is a surface made of metal or some other material placed in the journal where the turning shaft rides. A bushing often looks much like a wide metal ring with a hole or groove in it which allows lubrication to reduce friction. Because contact between two different metals reduces friction, the bearing and the shaft are usually different metals. The bearing is also a softer metal which when worn out can be replaced, providing protection for the shaft. Bushings, like roller bearings, are useful where the force of a turning shaft must be applied to a greater surface area.

Rollers and bushings reduce the friction between two surfaces to practically nothing. Design criteria determine which is appropriate for a particular application.

This author uses a beginning, middle, and end and states clearly what is being classified (bearings), what the basis is (design and purpose), and what the classifications are (roller and bushing).

For other student examples of division, study Models 1 and 2 in Chapter 6. Model 1 describes a stapler, a single mechanism, divided into three parts: main body, a plastic cover, and a base plate. Model 2 describes a ¾″ pipe elbow, a single mechanism. Even though the mechanism has no parts, the author divides it into parts or sections to discuss it: exterior portion between openings, the end pointing horizontally like the lens of a periscope, the interior, and the end facing downward.

CONCLUSION

Classification and division are fundamental approaches to organizing and structuring knowledge. As such, they will be part of most documents you write in college and the workplace.

REVIEW QUESTIONS

1. What is *classification?* Give an example.
2. What is a *genus?* Give an example.
3. What is a *species?*
4. Explain how genus and species relate. Give an example.
5. What is a *basis?* Give an example.
6. What is *division?* Give an example.
7. Is classification common in business and technical writing? Explain.
8. What is an easy way to distinguish *classification* from *division?*
9. What are some basic guidelines for writing classification?
10. What are some guidelines for writing division?

EXERCISES

1. Choose any five items from the following list. For each, draw a genus/species diagram and indicate the basis of classification.

Example: *Shirts*

Short Sleeve   Long Sleeve
Basis = Sleeve length

   a. Sandpaper
   b. Lamps
   c. Lawnmowers
   d. Motor Oils
   e. Books
   f. Periodicals
   g. Skis
   h. Stereos
   i. Records
   j. Telephones
   k. Light Bulbs
   l. Tires
   m. Radios
   n. Floppy Disks
   o. VCRs

2. Each level of an outline is much like a genus/species relationship. If the outline is to be logical, each level in each subsection must have a single common basis. Evaluate the following outline segments for a logical basis of classification. Identify any problems and correct them.

   a.   2.0 Reference books
                2.1 Dictionary
                2.2 Encyclopedia
                2.3 Atlas
                2.4 *Gone With The Wind*
                2.5 *Goodwin's Almanac*
   b.   4.0 Light Bulbs
                4.1 Globes
                4.2 Tubes
                4.3 Tear-Shape
                4.4 Incandescent
                4.5 Miscellaneous Shapes/Forms
   c.   3.0 Light Fixtures
                3.1 Fluorescent
                3.2 Floor
                3.3 Table
                3.4 Wall
                3.5 Ceiling

d.   6.0 Wood Species
            6.1 Eastern White Pine
            6.2 White Cedar
            6.3 Popular
            6.4 Hardwood
            6.5 Beech
e.   2.0 Abrasives
            2.1 Aluminum oxide
            2.2 Garnet carborundum
            2.3 Silicon carbide
            2.4 Coarse
            2.5 Medium
            2.6 Fine

3. Demonstrate your understanding of classification by writing a paragraph that classifies. Identify the subject, basis, and classifications (groups) in the topic sentence. Use the body of the paragraph to distinguish among the groups. Limit your classification to two or three groups.

4. Demonstrate your understanding of division by writing a paragraph based upon this pattern. State the subject and divisions in the topic sentence. Use the body of the paragraph to explain the divisions. Limit your subject to something that has no more than three or four parts.

5. *Major Writing Assignment.* Write a 500–750 word classification paper using three-part organization. In the introduction, clarify what is being classified, on what basis, and in what groups.

Choose a subject you already understand; avoid library research that may influence your wording and organization. Choosing a simple subject will let you concentrate on writing rather than trying to figure out the correct data. A subject with one to three bases and up to five classes is appropriate.

Because you can classify virtually any subject, the range of possible subjects is tremendous: cars, motorcycles, turntables, speakers (hi fi), books, music, movies, drugs, tools, saw blades, clouds, geological processes of earth shaping, razors, hair styles, clothing, and so on.

8
Process Description

Many reports, proposals, and other technical and business documents include a description of a process. In a technical society, writers and readers frequently need to explain, understand, and perform complicated technical processes. Process descriptions tend to be either machine-centered or operator-centered.

Because so much of industry is complex and highly automated, the role of the operator in the process is sometimes minimal; in fact, machine-centered process is more common. The Tree Milling Machine in Figure 8.1 is a good example. Controlled by a computer, it is capable of milling three-dimensional objects without human intervention.

A broader definition of machine-centered process would include natural processes such as the formation of clouds, tornadoes, tides, and a host of natural phenomena that are not performed or created by man. Machine-centered process, as defined in this text, is broad enough to include natural processes.

Operator-centered process, as the name implies, centers on processes involving an operator. Common applications might range from explaining how someone can repair a flat tire to flying a jet fighter. Operator-centered process may emphasize general understanding or actual performance of a process.

This first part of this chapter discusses writing a machine-centered process. The second discusses writing two basic forms of operator-centered processes:

Figure 8.1
Tree Milling Machine

the informative and instructive. The last sections contain writing models, review questions, and exercises.

WRITING MACHINE-CENTERED PROCESS DESCRIPTION

Writing machine-centered process description is similar to writing most documents. You must assess your audience, identify your specific purpose in writing for them, and design a document to meet their needs.

Your general purpose in writing is to describe how something works. Three-part organization and a minor modification of the standard four-part introduction are key elements of organization. At first you may find it awkward to write a description without an operator, but if you keep your purpose in mind this will not be a problem.

The next sections explain how to write the introduction, body, and conclusion.

Writing the Introduction

Most readers have questions in mind such as the following when they begin reading a process document:

- What are you talking about?
- Why are you talking about it?
- Why are you telling me about it?
- How does it work?
- What is an overview of the process?
- What background or theory do I need to start?

A variation of the standard four-part introduction (see Figure 8.2) can serve as an effective beginning. The amount of definition and background or theory will vary with the needs of the audience.

```
1.0 Introduction
    1.1 Subject (definition of terms if needed)
    1.2 Purpose
    1.3 Background/theory (as needed)
    1.4 Main steps/sequence of operation
```

Figure 8.2   Outline: Introduction

Section 1.4 introduces both scope and plan of organization.

Figure 8.3, an introduction to a machine-centered process description, contains most of the features mentioned in the outline given in Figure 8.2.

The turbo-charger, a common part of modern automotive engines, is a mechanism that pressurizes the mixture of air and gas before it enters the engine. Unlike the supercharger, which is belt-driven, the turbo-charger is spun by the pressure of exhaust gases. This paper describes the operation of one variation of the turbo-charger, called the "Comprex Supercharger." The description breaks the normally continuous and uninterrupted process into one revolution of the finned rotor, explaining its relationship to intake and exhaust.

Figure 8.3   Example: Introduction

Analysis of Figure 8.3 shows the subject is "turbo-chargers"; the purpose is to "describe the operation of one variation . . ."; the first sentence defines the term; the second provides important background; and the last suggests the scope and general pattern of organization.

Writing the Body

As with many writing situations, a good approach is to divide and explain. Divide the process into a sequence of operations that explain purpose and detail action. Clarify important cause/effect relationships with explanations such as this: "Exhaust gas entering through the port imparts pressure directly to the intake charge."

Figure 8.4 outlines typical organization for the body.

```
2.0 Description of Operation
    2.1 Sequence 1
        2.1.1 Purpose of the sequence
        2.1.2 Detailed description of the action

    2.2 Sequence 2
        2.2.1 Purpose of the sequence
        2.2.2 Detailed description of the action
```

Figure 8.4   Outline: Body

Writing the Conclusion

Because readers may have trouble pulling together all of the sequences of a process, the conclusion may serve for this purpose. An involved sequence may require a full paragraph to give an overview of all sequences discussed in the body (see Figure 8.5). A simple process, one the reader can easily keep in mind, may require little more than a single concluding sentence, for example, "This concludes the description of . . . ."

```
The overall operation of the Comprex Supercharger involves
a rotor which alternately pressures the intake mixture into
the intake port and then as the rotor passes the exhaust port
uses the exiting exhaust gases to pull in the next round of
intake air. This simple but effective mechanism produces sig-
nificant increases in the horsepower of an engine.
```

Figure 8.5   Example: Conclusion

STUDENT MODEL 1: MACHINE-CENTERED PROCESS DESCRIPTION

This paper is an author's response to an assignment to describe and explain how a simple mechanism or process works, emphasizing theory and operation. The purpose in the model is to give a detailed analysis of the parts and theory of operation of a marine water system. The audience, apparently, is a boat owner with average technical ability.

Model 1

MARINE WATER SYSTEM: PARTS AND THEORY OF OPERATION

The culinary water system in a small sail boat is simple in design and operation. It consists of an inlet, vent, storage tank, manual pump, and waste disposal outlet (see Figure 1). The purpose of this paper is to describe such a system and focus on its theory of operation.

Water enters the system through a 1-½″ flush deck fitting mounted somewhere above the internal storage tank. As water moves (gravity and pressure from hose) through plastic pipe to a 10 gallon storage tank, it displaces air already in the system—or the water won't flow properly. As the tank fills air vents from the system through a ½″ plastic line extended from the top of the

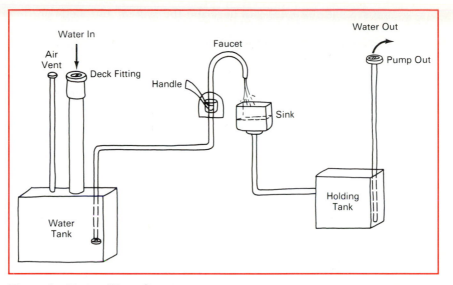

Figure 1   Marine Water System

storage tank to a deck opening located near the water intake fitting. When the tank is full, water flows from the air vent.

Operation of the system involves movement of water from the tank to the faucet and sink. The pump pulls water from the tank through an air-tight ½″ plastic line extending to the bottom of the tank. The manually operated pump creates a suction on the line with each stroke of the handle. Each stroke pulls a small plastic ball out of its fitting, sucking air from the line. The ball, acting as a small check valve, slips back into its housing instantly as suction begins to slip, blocking the line and maintaining most of the pressure or "prime" created by the first stroke. A second and third stroke create even more suction, usually sufficient to pull water into the line. Once water reaches the sink, approximately four strokes produce a cup of water.

Excess water leaves the sink through a gravity drain. A 1″ plastic pipe transports water to a holding tank, to avoid contaminating the water surrounding the boat. When the boat returns to the dock with a full tank, a shoreside facility pumps the waste water into a sewer system. A second deck plate provides an outlet for this purpose.

The components and theory of operation involved in a small boat water system are simple. Understanding them will make proper use and maintenance much easier for boat owners.

The model paper has many of the features discussed in this section. The introduction discusses the subject—"culinary water system in a small sail boat," states the purpose—"to describe the system and focus on theory of operation," and provides background in the second sentence by listing parts. No sequence of operation is listed, but for a short paper this is optional.

The body discusses three sequences: filling in the first paragraph, pumping or using the system in the second, and disposing of waste water in the third. Each sequence clarifies the purpose and gives details of the action.

The conclusion is short and to the point. The author uses strong, active verbs and avoids the passive. Because the paper emphasizes theory of operation and deliberately avoids mentioning an operator anywhere but in the last paragraph, it qualifies as a machine-centered process description.

WRITING OPERATOR-CENTERED PROCESS DESCRIPTION

Operator-centered process takes two basic forms: the informative and the instructive. The informative, sometimes called *theory of operation*, gives a broad overview of a process but does not include sufficient detail for the reader to perform it. By contrast, the instructive (the form characteristic of instruction manuals) is narrowly specific in detailing steps an operator takes to perform a process.

Informative process is useful to explain general theory. For example, readers of *Popular Mechanics* may want to understand the process involved in estimating automobile repairs. Their purpose in reading is not to become estimators, but to understand what happens at the repair shop when they take a car in.

On other occasions, these same readers may also want specific how-to-do-it information. A woodworker may want to make finger joints using a table saw but may lack experience and the appropriate jig. An instructive article will take the woodworker step by step through the process of constructing a jig and using it to make a finger joint.

WRITING INFORMATIVE PROCESS DESCRIPTION

The organization of an informative process description is similar to other documents; three-part organization and a four-part introduction are standard. Typical content aims at a broad overview; this may include background, relationship to other processes, evaluation of effectiveness, general theory and purpose, and overview of the process itself. In short, informative process includes everything a given audience needs if the readers are to gain a broad understanding of a process. Some highly detailed forms approach the function of the instructive but contain a broader range of information.

The subjects of *Equipment and Material*, so important in instructive processes, are sometimes included in informative processes as well. No standard rules exist about placement, but equipment and materials are often discussed immediately after the introduction and before an overview of the process. Generally the listing is not as complete as it is for instructive processes. For example, informative processes may mention "sandpaper," while the instructive specifies grits—80, 120, 220.

STUDENT MODEL 2:
INFORMATIVE PROCESS

This paper, representing half of Writing Assignment 8 in the exercises, functions very well as an informative process. It has three-part organization; the introduction clearly identifies the subject, purpose, and scope; and the content gives a broad overview of theory and process.

The author has also carefully controlled the verbs. The entire document uses active voice with occasional references to the operator—the car owner.

Model 2

FUSE: THE "WEAK" LINK IN A CAR'S
ELECTRICAL SYSTEM

A fuse, a small electrical device in an automotive electrical circuit, prevents excessive flow of current to components such as a radio, headlight, or heater. Understanding what a fuse is and how it works will allow a car owner to detect and possibly repair minor electrical problems.

A fuse (Figure 1) is a small cylinder varying in length from ½ to 1″. Both ends are metal caps connected to one another by a thin wire housed in a

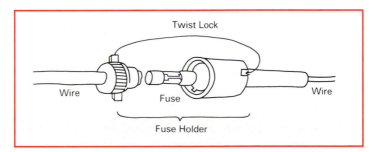

Figure 1   Fuse and Fuse Holder

transparent glass tube. The fuse, when placed in a circuit, allows a certain amount of electrical current to flow. Typical values are 5, 10, and 15 amperes. If the current flow exceeds the rating of the fuse, the wire inside the fuse melts, breaking the circuit, and protecting components from damage.

The owner of an automobile usually becomes aware of the fuses when the lights won't come on, or the radio, horn, heater, and/or windshield wipers won't work. A quick look at the fuse box, often located under the dashboard on the driver's side, will frequently reveal the problem. If one of the fuses, as viewed through the transparent section, is blown, the wire inside will have a clear break.

Sometimes replacing a fuse will solve a problem. Many times, a blown fuse indicates damage in a particular circuit requiring repair before a new fuse will help. If serious problems are present, the owner should take the car to a qualified repairman.

Fuses are extremely inexpensive when compared to the components they protect. For two or three dollars, a car owner can purchase a package of five and a tool to replace them. With a minimum of direction and experience, many owners can recognize and replace a blown fuse.

Understanding how a fuse functions as a carefully engineered "weak link" may help a car owner understand electrical problems—and even repair them.

WRITING INSTRUCTIVE
PROCESS DESCRIPTION

Instructive process (performance instruction) differs from informative process in purpose and format. Figure 8.6 illustrates the differing purposes of the two forms of operator-centered process.

Instructive process is one of the most common writing tasks in business and industry: every new product needs an instruction manual; every procedure must have a standard reference manual.

Purpose

Instructive process involves a series of steps to tell the reader how to do something. Because the focus is so much narrower than in informative process, the information must be limited to data the reader needs to perform the process.

Format

To create an *introduction*, use a variation of the standard four-part opening: present the subject and purpose, suggest the scope, and provide minimal background. In addition, it is often useful to identify the intended audience and

Figure 8.6
Relationship: Informative and
Instructive Process

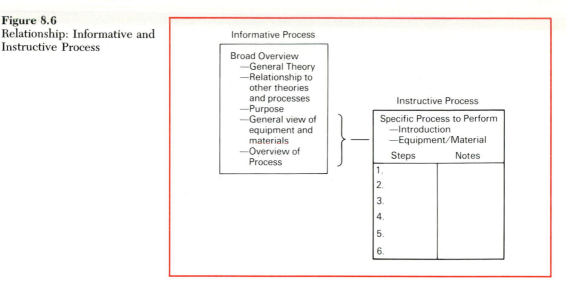

skill level and, when appropriate, warn of possible hazards associated with the process.

The format of instructive process usually includes a *listing of parts and equipment* needed for performing the process. Logically, you may place this information immediately before the step-by-step instructions or in the body of the paper as needed in the process. In a one-to-three page document, an initial listing will probably be easy for the reader to recall. In a longer document, however, it may be more helpful to place parts and equipment sublists near each sequence of steps.

Make the parts and equipment list specific. For example, *cleaner* could mean many different solutions and powders; a *3-1 acetone wash solution* is a much more helpful identification. If your reader will need a screwdriver, specific the type: regular or Phillips.

In the *step-by-step instructions*, break the process into a series of detailed steps written exclusively in the *imperative* or *command form of the verb*. For example, if you want the reader to open a valve to allow water to flow, say "Open valve #12 approximately three turns until the meter reads 45 lbs. pressure." If you want the reader to wash something, say "Wash each silver-chloride disk thoroughly by squirting it with the acetone wash solution." The imperative verbs *open* and *wash* have the authority of a command.

Imperative verbs are economical and may help you avoid an awkward repetition of the word *subject* or *operator*. For example, instead of saying "The operator should open the valve. . . ," use the imperative "Open the valve. . . ," a savings of three words. If you have a series of steps and have to repeat the word *operator* in each of them, the repetition becomes monotonous and awkward.

The natural way to solve the problem of repetition is to use a pronoun to replace *operator;* however, this may lead to other problems. Which pronoun should you use—*he* or *she?* The imperative avoids potential sexist language and the awkwardness of saying *he/she* for each step. As you write instructive process description, reserve the imperative for steps the reader must perform.

Many times you will want to supplement a step with additional information or a note such as is shown in Figure 8.7.

```
Step 5.
    Place second disk (indented side up) on top of the first
    disk and sample.
        Note: There should be no air bubbles between disks.
              If bubbles persist, step four may be repeated
              with two drops.
```

Figure 8.7

Keep steps and notes physically separated on the page by using an indentation (as in Figure 8.7); by using two columns—the first for steps, the second for notes (see Model 3, p. 124); by using different typefaces or other methods. Separation highlights steps and enables the reader to avoid confusion.

In writing notes, use active or passive verbs. If you use the imperative in notes, they will appear to be steps, and may be a possible source of confusion to the reader. Figure 8.8 illustrates differences among the imperative, the active, and the passive.

```
Imperative:  Cut the plate 2" from the end.
Active:  The operator cuts the plate 2" from the end.
Passive:  The plate is cut 2" from the end by the operator.
```

Figure 8.8   Verb Forms Compared

Chapter 15 of this text includes an extensive discussion of the active, passive, and imperative forms.

USING WARNINGS, CAUTIONS, AND NOTES

Instructions often contain procedures that could cause loss of life or injury, damage to equipment, poor results, or other problems. Although many manuals may feature an initial safety requirements section, it is important that this information also appear in the process, usually immediately before a specific step.

Here are the most frequently used precautionary and information notes:

Danger	Apply this term to any procedure or operation that could lead to loss of life or serious injury. Use this label when you make reference to flammable or poisonous materials or high voltages.
Warning	Use with steps where damage could occur to equipment and materials.
Caution	Use with steps where poor results may occur.
Note	Use anytime to point out potential problems or to give extra information.

Of course, not everyone will define these terms in the same way. If you were writing a long manual, it would be good practice to explain your interpretation by defining your use of the terms in the preface or introduction.

Practice varies about which verb form to use in notes. One air force flight manual contains the following explanation:

The words "shall" or "will" are to be used to indicate a mandatory requirement. The word "should" is to be used to indicate a nonmandatory desire or preferred method of accomplishment. The word "may" is used to indicate an acceptable or suggested means of accomplishment.

Courtesy Department of Defense

These distinctions are generally representative of common practice.

Warnings and cautions should be placed in a conspicuous location before a potentially hazardous step. Use large letters, contrasting typefaces and ink colors. These will draw the attention of readers. At times, drawing a box around the warning is effective.

STUDENT MODEL 3: INSTRUCTIVE PROCESS

The author of this paper is responding to an assignment to write two short (approximately 250 words) documents, one featuring the broad general view of informative process (Model 2), and another featuring the narrow, specific view of instructive process (Model 3). Although both discuss the same general topic, they are independent of one another.

Model 3 focuses on one specific aspect of "Fuse: The 'Weak' Link in Your Car": it explains in a step-by-step process how to replace a blown fuse. As you read, notice how it differs from Model 2 in range of subject matter, the focus of the introduction, the specific parts list, verbs, and careful separation of steps and notes.

Model 3   REPLACING A "BLOWN" FUSE

A fuse is an inexpensive device used to protect electrical components in a car from excess current. A fuse "blows" or melts before current can damage a component such as a radio or heater motor. These instructions explain how to locate the fuse box, identify the bad fuse, and replace it. The entire process requires little mechanical knowledge and is within the ability of the average driver.

Tools and Materials

- 1 replacement fuse with proper current rating
- 1 fuse puller
- owner's manual for the car

Procedure

STEPS	NOTES
1. Locate the fuse box.	1. The owner's manual will show the location. If a manual isn't available, the three most common locations are under the dash on the driver's side, under the hood, or in the glove box.
2. Identify the bad fuse by inspecting the fuse box (see Figure 1).	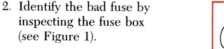 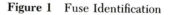 **Figure 1** Fuse Identification
3. Clamp the fuse puller around the middle of the fuse and pull outward.	3. The fuse may require a strong pull to release it.
4. Verify the rating of the old fuse and replace it with one of the same rating.	4. Many fuse panels have the rating printed above the fuse. The rating is also written on the fuse. Most range from 5–15 amp.
5. Place a new fuse in the puller.	5. The middle of the fuse should go in the puller.

6. Using the tool, push the
   new fuse into fusebox.

7. Verify it is in tightly.   7. A loose fuse won't work properly.

The circuit is now ready for testing.

Model 3 uses imperative verbs in each of the steps. The notes are in both the active and passive voices. This document is short, but instruction manuals for complicated processes can occupy hundreds of pages.

CONCLUSION

Process description, either machine-centered or operator-centered, is a common part of writing on the job. Machine-centered process, broadly defined, may include descriptions of natural processes such as the formation of clouds. Operator-centered process varies in purpose. Sometimes its aim is to give a broad overview of a process performed by an individual; sometimes its purpose is to give specific instructions so an operator can follow a series of steps to complete a process.

REVIEW QUESTIONS

1. Which is more common, operator or machine-centered process description? Why?
2. What is the typical content for the introduction of a machine-oriented process description?
3. How is the body organized?
4. What is the basic function of the conclusion?
5. What are the two common forms of operator-centered processes?
6. How do the two forms differ in purpose?
7. How do they differ in content and format?
8. Which form uses imperative verbs?
9. Why is it important to keep steps and notes clearly separated in instructive processes?
10. Where should warning and caution notes be placed in instructive processes?

EXERCISES

1. Write one or two paragraphs describing the operation of a mechanism as a machine-centered process. Do not include an operator. Choose something familiar such as a bicycle, mechanical pen, pencil sharpener, or balance scale.

2. Find an article in a magazine that contains a description of machine-centered process. Write a short analysis and submit it with a photocopy of the article.

3. Write a short instructive procedure explaining a simple process such as putting lead in an automatic pencil, making a backup disk on a computer, or changing the ball on an electric typewriter. Include numbered "steps" written in the imperative and "notes" written in the active and/or passive. Keep steps physically separated from notes.

4. Analyze the owner's manual for a small electrical appliance or tool such as a string trimmer or electric drill. Include the following: What kinds of information are included? Are the instructions for operation clearly written? What verb forms are used? Are steps and notes clearly separated? Are warning or danger notes placed in useful positions?

5. Analyze the following instructions for logical patterns of verb usage; particularly compare part **a** with part **b**. Now revise to make the two paragraphs consistent in format and use of the imperative.

   a. Handle Panel Attachment.   The handle panel is attached by sliding it down over the handle brackets on the chassis and installing the four carriage bolts in the lower holes of the handle panel. The bolts will be found in the hardware bag. The handle height may be changed by loosening the four bolts which attach the lower handle brackets to the chassis, moving the handles to the desired position and re-tightening all the bolts. The handles should be set slightly above waist level.

   b. Throttle Control Attachment.   Route the throttle control cable inside the lower right handle bracket and up to attach to the carburetor. Insert the end of the cable wire into the carburetor control arm and place the cable under the clamp as shown below. Move the throttle control on the handle bar to the full choke position and place the carburetor control arm in the full choke position. Tighten the control handle and the carburetor arm should move freely.

6. Rewrite the instructions included in Exercise 5 as a series of numbered steps and notes. Which form is easier to use?

7. Write a 500-word paper emphasizing a machine-centered process. Base the organization of your paper on Figures 8.2 and 8.4 with a review of Model 1. Use active verbs and support the text with appropriate visuals.
   Choose a process you know well, perhaps one you studied in another class. Appropriate subjects might include how clouds are formed, how an electric pencil sharpener works, how a sewing machine links stitches, how food is digested, how a heart pumps blood, and so on. Whatever you choose, avoid emphasizing an operator in the process.

8. Write two short operator-centered process papers of 250–350 words each on a common topic. Make the first part an informative process, a broad view of the subject written in standard three-part organization (see Model 2). Make the second part an instructive process, explaining through a series of steps how to do something (see Model 3). Review Figure 8.7 for a clarification of the relationship of the two papers.
   This writing assignment provides a good opportunity for you to demon-

strate your control of verb forms. Write the informative process with either active or passive verbs, preferably active. In the instructive process, use the imperative for steps to perform and the active or passive in the notes. Keep steps and notes physically separated. Try to use appropriate visual support in both sections of the paper.

Turn the paper in under one cover sheet with the informative process first.

9

Interpretation of Data

A primary activity of business managers, scientists, and engineers is to gather and intepret data. Managers collect data to study before investing heavily in new equipment, increasing inventory of supplies, or buying another business. Scientists perform an experiment or gather data by observing some phenomenon; however, the lists of numbers and observations mean nothing until interpreted.

You interpret data nearly every minute of your life. As you walk down the crowded halls of a building and attempt to pass someone, you mentally record the speed of the other person, his or her position in the hallway, any body-signals that may indicate a left or right turn, and the relative position of anyone approaching. From all this data you reach the decision to pass or to wait for a better opportunity.

When you want to buy something, you probably do comparison shopping. For example, if you want to buy a new SLR35 mm camera, you examine several models in your proposed price range, compare their features (shutter speeds, lens quality, and so on), check several stores for price, and then make your decision.

Interpretation of data is the end step of a process that begins with collecting data. Once you have collected it, you have to analyze it, organize it in some way so you can understand it, and, if necessary, present it. In business, interpretation of data may be the basis for making a recommendation or decision; in basic research, interpretation of data may be just one more step in the long process of understanding a subject.

This chapter discusses basic guidelines for working with data, then explains how to write the introduction, how to organize the body, how to write conclusions and recommendations, and how to use graphic support. The chapter concludes with models of student writing, review questions, and exercises.

WORKING WITH DATA

Objectivity is the goal of most business and scientific writing. For example, if you are comparing cameras, you may develop a personal preference or bias for a particular unit simply because you like its shape or color even though it costs more and has no extra features and may even have fewer. If this is the situation, your emotions—not objectivity—influence the decision.

Technical writing requires a neutral scientific attitude. Interpretation should fit within the limitations of what the data can support. You should avoid inferences and personal opinions that go beyond the data. If you feel compelled to state your inferences and personal opinions, tell the reader that is what they are!

Use valid data to support your recommendations. For example, if you recommend replacing a piece of equipment to gain greater economy and efficiency, be certain of your facts. Will your company's production schedule allow a level of production sufficient to gain economy and offset the cost of new equipment?

What has been the experience of other companies that made the change? Even though the machine succeeded in another company, will variables produce different results in your firm?

When you compare several alternatives, be careful about your conclusions. For example, if the first of two alternatives is obviously bad, this does not mean that the second alternative is automatically good. It may be bad as well! Examine alternatives carefully and objectively before making a recommendation.

Depending upon the type of research done, you may have a considerable volume of data available when you write. As a general guideline, in the body of the document use only the data directly related to the discussion. Place raw data or additional supportive data in the appendix or exclude it altogether. Considerations of audience and purpose will help clarify such decisions.

WRITING THE INTRODUCTION

Although interpretation of data often appears as part of a longer document, it is also common in short feasibility or recommendation reports. The type of introduction to use is a variation on the standard form mentioned in Chapter 3. It should contain information on the subject, purpose, source of data, scope and limitations, and a statement of organization and criteria.

Examine Figure 9.1 to see how these features appear in an introduction.

PUMP SHOTGUNS

Pump shotguns, according to those who use them, are a very effective firearm in slaying waterfowl, upland game birds and even big game like deer and antelope at ranges up to 100 yards. With so many pump shotguns on the market, choosing the best one is difficult. Within the past several years two models, the XXX 2000 and the XXX 870, have become very popular. Based upon literature from each company, this report will compare the two shotguns on the basis of (1) price, (2) durability, (3) flexibility, and (4) capability.

Figure 9.1 Sample Introduction

Here is an analysis of the five features included in Figure 9.1:

- Subject  . . . . pump shotguns
- Purpose . . . . evaluate two popular models

- Source of Data . . . . . company literature
- Scope/Limitation . . . . . two most popular models
- Statement of Organization . . price, durability, flexibility, capability

The author assumes an audience interested in guns and hunting. The paper results from the author's interest in evaluating two popular shotguns.

Although the introduction illustrated by Figure 9.1 does not fully indicate the particular approach, this paper uses criteria-based organization; that is, as the statement of organization indicates, *price* is discussed first, followed by *durability*, *flexibility*, and *capability*. Both guns are discussed in terms of each criteria.

Readers are often interested in why a writer limits the scope in a particular way and in where the author obtained the data. If, for example, you base your paper on material from a periodical such as *Consumers Report*, June 1986, p. 10, explain this to your readers. If you have multiple sources, a generalization in the introduction supported by a list of works cited at the end of the paper may help readers evaluate the accuracy of your interpretation.

ORGANIZING THE BODY

Once you have written the introduction, with the statement of organization/ criteria appearing last, you have also established a pattern for the body. Two common organization patterns are particularly useful: (1) item-based, and (2) criterion-based.

Item-Based Organization

List items you want to discuss in the introduction and then discuss them one at a time, applying all criteria to each one in turn. Figure 9.2 features an outline based upon this pattern.

Such a plan is especially workable when you are comparing only two or three items. It is also very easy to write; you simply tell all about one item and then all about the next, and so on. Such an arrangement has limitations for readers, however, because they must remember what was said about Item 1 and Criteria 2 as they read about Item 2 and Criteria 2. As long as the comparison is short and simple, this is not a serious problem.

A practical example of this format is illustrated by a comparison of several small foreign cars—Volkswagen, Datsun, and Toyota. The cars would be compared in terms of miles per gallon, maintenance costs, and initial cost. In the item-based format, the author would discuss the Volkswagen in terms of these criteria; then the Datsun in terms of the same three criteria; and then the Toyota.

```
1.0 Introduction
    1.1 Subject
    1.2 Purpose
    1.3 Source of Data
    1.4 Scope/limitations
    1.5 Plan-of-development (items and criteria)

2.0 Body
    2.1 Item 1
        2.1.1 Criteria 1
        2.1.2 Criteria 2
        2.1.3 Criteria 3
    2.2 Item 2
        2.2.1 Criteria 1
        2.2.2 Criteria 2
        2.2.3 Criteria 3

3.0 Conclusion/Recommendatins
```

Figure 9.2   Outline: Item-Based Organization

Criterion-Based Organization

Instead of telling all about one item and then all about the other, you can organize your writing around the criteria. List a criterion—miles per gallon, for example, and then discuss the three cars—but only in terms of miles per gallon. Then move on to the next criterion.

Figure 9.3 is an example of criterion-based organization. The general introduction to both patterns is similar.

The criteria-based plan demands more of the writer but often produces greater clarity for the reader, especially if several items with several criteria are compared. Readers do not need to recall data for comparison because it is presented directly side-by-side. This arrangement also lends itself readily to graphic presentation.

WRITING CONCLUSIONS AND RECOMMENDATIONS

In many papers that interpret data, you can draw conclusions in the body of the paper as you consider each of the criteria; this is a common and worthwhile practice. The conclusion to such a document is typically a short paragraph containing a summary statement for each criterion applied. For example, if the body

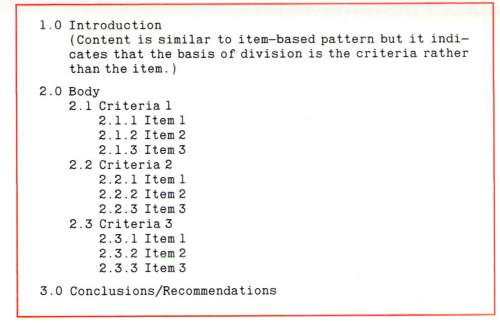

```
1.0 Introduction
    (Content is similar to item-based pattern but it indi-
    cates that the basis of division is the criteria rather
    than the item.)

2.0 Body
    2.1 Criteria 1
        2.1.1 Item 1
        2.1.2 Item 2
        2.1.3 Item 3
    2.2 Criteria 2
        2.2.1 Item 1
        2.2.2 Item 2
        2.2.3 Item 3
    2.3 Criteria 3
        2.3.1 Item 1
        2.3.2 Item 2
        2.3.3 Item 3

3.0 Conclusions/Recommendations
```

Figure 9.3 Outline: Criterion-Based Organization

of your paper covers four criteria, then use four summarizing statements—major points only—as the basis for the conclusion. Notice how summary and recommendation function in Figure 9.4.

In shorter papers (one–four pages) you can easily combine the conclusions (summary) and the recommendations in one paragraph. Many times the conclusion is already so apparent to the reader that you need spend little time and space writing it.

```
    The best method of depreciation depends on many factors.
For a new business just starting out the straight-line method
is the best to use because it gives the greatest book value,
therefore showing more profit for the business in its first few
years. For a well-established business the declining-balance
method is best because it calculates higher depreciation in
the early years, causing the book value to decrease rapidly
in the first few years.
```

Figure 9.4 Example: Conclusion

USING GRAPHIC SUPPORT

Papers that interpret data particularly need graphic support. A heavy mixing of numerical data with explanations is difficult to read because readers have to group the data mentally. Numbers are usually easier to handle in columns or groups of some kind. In Figure 9.5, compare the following passages that are written in two ways?

```
Version 1:   The following information shows how each collec-
tor performed on a summer day of 70 degrees average tempera-
ture: Collector 1, 145 degrees; Collector 2, 130 degrees;
Collector 3, 135 degrees: Collector 4, 115 degrees.

Version 2:   The following information (based on summer days;
70 degrees average temperature) shows Collector 1 is the most
efficient because it heated water 10 degrees warmer than the
next best unit:

Collector 1        145
Collector 2        130
Collector 3        135
Collector 4        115
```

Figure 9.5   Comparison: Data Presented Two Ways

The second version is more helpful to the reader because it makes it easier to compare the figures. This version also illustrates the proper relationship between the graphic data and the explanation. It is not necessary to restate in the body of the paper everything mentioned in a table or visual. On the other hand, it is necessary to *refer* to the graphic material in the written text and to provide enough interpretation so that the reader can easily understand the data.

Another consideration in using graphics is *placement*. If you have a page of data, you may be tempted to group it all on one page and simply ask the reader to turn to it when you want to refer to it. This is a handy arrangement in some ways, particularly because all the data is together, enabling the reader to see relationships. However, a page of data can become formidable and confusing to the reader.

In the criteria-based approach, it is useful to include the comparative data for just one point in a small box immediately adjacent to the discussion. At the end of the paper, in the conclusion, it may be useful to assemble all the little bits of graphic data into one chart or illustration, particularly in a longer paper that uses considerable data. For a complete discussion of graphic or visual support, review Chapter 4.

STUDENT WRITING MODELS

Both of the following writing models use the suggested format for writing the introduction, body, and conclusion; their use of graphics is helpful to the reader; and in each case the authors clearly structured and organized the writing themselves, avoiding any excess influence from their sources of data.

In Model 1 the author credits *Consumers Report* as the source of data. To make the data fit his paper, and the number of drills evaluated he reduced the number of criteria used in the original article from six to four. Eliminating the top and bottom rated drills changed the outcome of the original study and allowed him to interpret the data himself, decreasing his dependence on the text of the original article. If you are not doing original research, this is a good approach to gathering data, because it allows you to concentrate on writing.

MODEL 1   **INTERPRETATION OF DATA**

In order to connect something with bolts, screws, anchors, rivets, or dowels, one often must drill a hole. For this reason the electric drill is usually the first power tool purchased for the home. But of all the types and brands, which will be the best suited for household use? The purpose of this paper is to evaluate the performance of specific models commonly available to a homeowner.

Four drills were chosen for testing: Black and Decker 7190, Sears 1148, Sears 1004, and Skil 457. Drills chosen had the following features: variable speed; double insulation, which allows safe use with two-prong power cords; a ⅜ inch drill chuck, the most popular size for household use; price range between $35–$75.

The criteria used in the evaluation were sustained power output, ability to drive screws, handle comfort, and price. All tests and results were taken from the May 1982 *Consumers Report*, pp. 262–265.

Sustained Power Output

Sustained power output, an indicator of the drill's ability to perform for a prolonged period under a load without overheating, was measured by running each drill on the dynamometer against a one-foot-pound load for three 10-minute periods. After each period, the drill's maximum power output was measured.

Most of the drills did not suffer a significant loss of power. The Black & Decker 7190 and Sear 1148 were the best in this test.

Sustained Power Output

	EXCELLENT	VERY GOOD	GOOD	FAIR	POOR
B & D 7190		X			
Sears 1148		X			
Sears 1004			X		
Skil 457			X		

Ability to Drive Screws

The ability to drive screws was a test designed to check the variable speed control and the torque of the drills. To see how well each drill could cope with these two requirements, hundreds of screws were driven into a 1½ inch thick fir board and then removed, using the reversing feature.

The Sears 1004 was rated fair because the variable speed was not smooth which made it difficult to drive screws. The Sears 1148 received the best rating at *very good*.

Driving Screws

	EXCELLENT	VERY GOOD	GOOD	FAIR	POOR
B & D 7190			X		
Sears 1148		X			
Sears 1004				X	
Skil 457			X		

Handle Comfort

Handle comfort, along with good balance, can make a difference if the drill is used frequently. The Skil 457 was judged fair because its handle was so short that you had to grasp it high, with the middle finger instead of the index finger on the trigger. The Black and Decker 7190 was also judged fair because its squared corners dig into the palm of your hand. The best of the six was the Sears 1148 which had an extra long handle that was very comfortable to hold.

Handle Comfort

	EXCELLENT	VERY GOOD	GOOD	FAIR	POOR
B & D 7190				X	
Sears 1148		X			
Sears 1004			X		
Skil 457				X	

Price

A big concern of many consumers is price. If the drill is going to be used often, it may be wise to spend more and receive one that is more durable. The most important thing is to get the best value for the money. The Sears 1148 was the most expensive and the Skil 457 the least expensive.

Prices

B & D 7190	35$
Sears 1148	75$
Sears 1004	35$
Skil 457	53$

In this paper four drills were judged against four standards: sustained power output, ability to drive screws, handle comfort, and price. The Black & Decker 7190 and the Sears 1148 are similar in performance but differ significantly in price. Because of its good performance at a reasonable price, the Black & Decker 7190 is rated the best choice for a household drill.

Model 2 assumes an audience of consumers with limited knowledge of 35mm cameras. The discussions of ASA Range, warranty, and exposure modes include careful definition of basic terms.

MODEL 2   CHOOSING A 35MM CAMERA

Today many people are turning to the 35mm Single Lens Reflex (SLR) cameras to meet their photographic needs. With many different cameras available, consumers need to find the best values. The purpose of this paper is to evaluate four popular models of 35mm SLR camera based upon their ASA range, warranty, and explosure modes.

The four cameras compared are the Canon AE-1 Program, Minolta X-700, Pentax ME Super SE, and the Konica FP-1 Program. All are in the $400–500 price range sold with body, lens, and shoulder strap. Information used in the comparison came from the March 1983 *Consumer's Research*, pp. 11–16.

ASA Range

ASA range refers to the variety of film speeds a camera will accept. The rate which light reacts to film is "film speed." Film speed is calibrated in numbers set up by the American Standards Association (ASA). A camera with a wide film speed range is desirable because of the flexibility it allows in choosing films. Table 1 lists the four cameras with their film speed ranges.

Table 1

CAMERA	ASA RANGE
Canon AE-1 Program	12–3200
Minolta X-700	25–1600
Pentax ME Super SE	12–1600
Konica FP-1 Program	50–400

The Canon AE-1 Program has the widest, and most versatile ASA range.

Warranty

Before purchasing a camera, the consumer should carefully check the warranty. If something does go wrong, it is important to have a good warranty. Most cameras have warranties, but they are not all the same. Table 2 lists the four cameras and their warranties.

Table 2

CAMERA	WARRANTY
Canon AE-1 Program	"Limited" one year against defective material and workmanship.
Minolta X-700	Two years against defects in material or workmanship.
Pentax ME Super SE	"Limited" two years against defects in material or workmanship.
Konica FP-1 Program	"Limited" one year against defects in material or workmanship.

The Minolta X-700 has the best warranty, with two years coverage for defects in materials and workmanship.

Exposure Modes

Many photographers like the versatility of having more than one exposure mode. An exposure mode is a system for controlling one to two of the possible variables in exposing the film. There are four common exposure modes: manual, shutter speed priority, aperture priority, and program.

In "manual mode" the operator controls both shutter speed and the aperture. In "shutter speed" priority, the operator sets the aperture and the camera controls shutter speed. In "aperture priority," the operator sets the shutter speed, and the camera controls the aperture. In "program mode," the camera is fully automatic. Table 3 lists the four cameras and their exposure modes

Recommendation

Overall the Canon AE-1 rated "best" in two of three categories and is, therefore, the best value of the four cameras. However, if the consumer values a good warranty, and does not require four modes of exposure, the Minolta X-700, which rated second best in this evaluation, is also a good value.

Table 3

CAMERA	MANUAL	SHUTTER SPEED PRIORITY	APERTURE PRIORITY	PROGRAM
Canon AE-1 Program	X	X	X	X
Minolta X-700	X		X	X
Pentax ME Super SE	X	X	X	
Konica FP-1 Program	X			X

The Canon AE-1 Program has four modes, making it the most versatile.

Model 2 has three-part organization, a solid introduction (first two paragraphs), clear headings, effective in-text summaries, and a clear conclusion. This paper, in contrast to Model 1, uses active verbs. Compare verb usage in the two papers and decide which you feel is more effective.

CONCLUSION

Objectively handling data as a basis for making informed decisions for developing an understanding of a subject is a common activity in business, technical, and scientific writing. This chapter, Interpretation of Data, is the last in Part II–Writing Basic Patterns. The next section, Part III, discusses writing business correspondence.

REVIEW QUESTIONS

1. Is interpretation of data a common activity? Give an example.
2. What is a scientific attitude toward data?
3. What is the general rule concerning the placing of data in the body of the paper? When should it appear in the appendix or not at all?
4. How is the introduction to a paper that interprets data similar and different from a standard four-part opening?
5. What are *item-based* and *criterion-based* organization?
6. What are the strengths and weaknesses of the two approaches?

7. What is typical form for a conclusion?

8. What does this book suggest concerning visual support for a paper that interprets data?

EXERCISES

1. Find examples of periodical articles or other documents based upon interpretation of data. One common type is a product comparison and evaluation. Analyze the approach used:
   a. assessing feasibility—will something work in a given situation?
   b. comparing two or more items—such as in the student writing models and in consumer product evaluations.
   c. solving a problem—why did something happen?
   d. answering a practical question—will something work for a given purpose?
   After identifying the approach used in interpreting the data, write a one-page descriptive summary of the article, and include accurate bibliographical information.

2. Demonstrate your understanding of item-based analysis by writing a paragraph using this pattern. Limit content to two items and two or three criteria.

3. Demonstrate your understanding of criteria-based analysis by writing a paragraph using this pattern. If you have written Exercise 2, rewrite it to fit this pattern. Limit content to two or three criteria and two items.

4. Write a 500–750-word paper interpreting data related to a research subject, major, hobby, or similar subject. Your approach should be to recommend a certain course of action as opposed to a simple explanation of data. Use a criteria-based (or item-based) organizational pattern. Place supporting data in graphic or tabular form when possible and use the writing to point out significant relationships. State conclusions as you reach them in the body of the paper even though you may need to summarize them at the end. Follow the suggestions on pp. 130–132 for writing the introduction, body, and conclusion.

   If you are not doing a research project, it is appropriate to borrow data from a source such as *Consumer's Report* or some other publication that compares products and data as a basis for making a selection. Many articles have excess data for a paper of the length suggested. Select a limited number of criteria from your source—i.e., three or four—and disregard the rest. This will usually give you an opportunity to reinterpret the data and draw your own conclusions.

   Do not let the source influence your writing—use it as a source of data and write your own interpretation using the patterns specified for this assignment.

PART III

WRITING BUSINESS CORRESPONDENCE

10

Business Letters

Writing letters is one of the most common—and most important—duties for people employed by business and industry. Advertising, buying and selling goods, dealing with government regulations, answering complaints, and making inquiries are a few of hundreds of ways business relies upon written correspondence.

In a very real way, every letter that leaves an office is a public relations document representing its author and employer. Carefully written letters can increase sales and create a favorable impression, just as poorly written letters can lower sales and reduce public trust.

Rather than attempting to deal specifically with the hundreds of letter-writing situations possible, this chapter focuses on fundamental principles that apply to all business correspondence. It discusses format, parts of letters, tone and word choice, organization, and some basic types of correspondence. This is followed by a letter of application and résumé, and by review questions and writing assignments.

FORMAT

Writing effective letters involves understanding and following conventions of format. These conventions are so well known that most readers have come to expect them; in fact, some may question your competence as a writer if you do not follow them.

Business writers generally use one of four common formats: block, modified block, full block, or simplified block.

- Block

The inside address, salutation, and paragraphs are placed at the left margin. The return address, complimentary closing, and signature start at or just past the middle of the page, toward the right.

- Modified Block

This format is identical with the block except for paragraph indentations.

- Full Block

All parts of the letter are placed at the left margin. With this format, some writers do not use a colon after the salutation and a comma after the complimentary closing.

- Simplified Block

Similar to the full block format, this version deletes the salutation and complimentary closing. This format is the least used of the four, although it has gained increased acceptance in recent years.

See Figure 10.1 for a comparison of the four forms.

The general appearance of a letter is also important. Besides being neatly typed, the letter should have enough white space to invite the eye and should follow these minimum standards for spacing and margins:

Figure 10.1
Letter Layout Formats

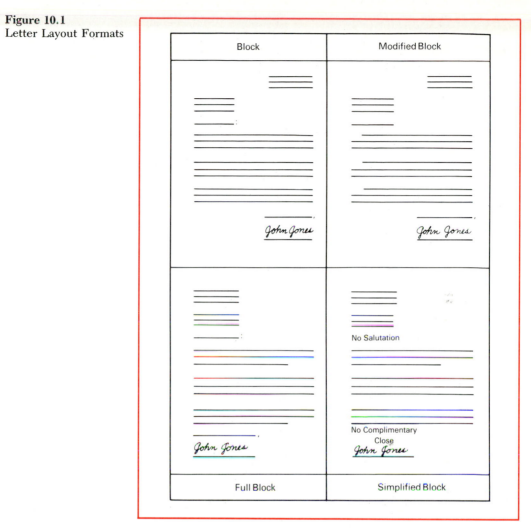

1. Allow a minimum 1½-inch margin at the top and bottom of the page.
2. Allow a minimum 1-inch margin on the sides.
3. For short letters, increase the margins and center attractively on the page.
5. Single space within the parts of a letter and double space between the parts.
6. Retype the letter if erasures or other unattractive features damage a favorable impression.

Figure 10.2 shows typical spacing.

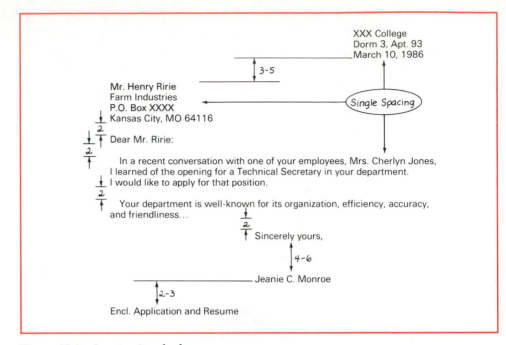

Figure 10.2   Spacing Standards

PARTS

This section discusses the eight common parts of a business letter: heading, inside address, attention and subject lines, salutation, body, complimentary closing, signature, and optional data lines.

Heading

Include your complete mailing address and the date, but not your name. Place the heading at the top of the page with the longest line ending at the right margin. If you use letterhead stationary, add only the date.

Example: Macmillan, Inc.
866 Third Avenue
New York, NY 10022
January 21, 1986 (or 21 January 1986)

Although you should avoid most abbreviations in the heading and inside address, it is appropriate to use two-letter abbreviations for states. Notice they are all capital letters, not followed by periods.

State two-Letter Abbreviations

Alabama	AL	Montana	MT
Alaska	AK	Nebraska	NB
Arizona	AZ	Nevada	NV
Arkansas	AR	New Hampshire	NH
California	CA	New Jersey	NJ
Colorado	CO	New Mexico	NM
Connecticut	CT	New York	NY
Delaware	DE	North Carolina	NC
District of Columbia	DC	North Dakota	ND
Florida	FL	Ohio	OH
Georgia	GA	Oklahoma	OK
Hawaii	HI	Oregon	OR
Idaho	ID	Pennsylvania	PA
Illinois	IL	Rhode Island	RI
Indiana	IN	South Carolina	SC
Iowa	IA	South Dakota	SD
Kansas	KS	Tennessee	TN
Kentucky	KY	Texas	TX
Lousiana	LA	Utah	UT
Maine	ME	Vermont	VT
Maryland	MD	Virginia	VI
Massachusetts	MA	Washington	WA
Michigan	MI	West Virginia	WV
Minnesota	MN	Wisconsin	WI
Mississippi	MS	Wyoming	WY
Missouri	MO		

Inside Address

Place inside address at the left margin and at least two spaces below the heading, followed by this information:

- Full name of the person addressed, including titles such as Mr., Ms., Dr., and so on.
- Name of company written in the same form the company itself uses.
- Complete mailing address. In street addresses avoid abbreviations.

Examples: Martin Bird & Associates
326 First Street
Annapolis, MD 21403

Mr. Mike Pyzel
Pyzel Navigation & Cruising School
86 Olive Mill Road
Santa Barbara, CA 93108

Attention and Subject Lines

An *attention line* is used when writing to a company rather than an individual. Write ATTENTION in capital letters (or abbreviate—Attn.) followed by a colon and the name of the department, office, or individual. Place this data at the left margin, two spaces below the inside address.

Example: ATTENTION: Accounting Office

or

Attn: Accounting Office

A *subject* (Reference) *line* is used to clarify key information early in a letter. Write "Subject" or "Re" followed by a colon and specific information such as letter, file, policy, or contract number.

Example: Subject: Specification No. 2BV-237

Location of the subject line varies widely. Many writers place it at the left margin, two spaces below the inside address.

Salutation

Place the salutation two spaces below the inside address (or attention/subject line) at the left margin, and follow it with a colon. The saluation usually begins with "Dear," followed by a title of respect plus the person's last name.

Example: Dear Ms. Smith:

Traditionally many letters have been addressed to "Dear Sir" or "Gentlemen" if the reader's name is not known. These greetings ignore the possibility that a woman will receive the letter. No sex-neutral greeting is standard in business correspondence. One reasonable solution is to *Dear Sir or Madam.*

Body

Begin the body of the letter two spaces below the salutation. Single space within paragraphs and double space between paragraphs. Variations in other features such as indentation depend on the format you select for the letter.

If your letter continues to a second page, write a heading containing the name of the person you are writing to, the page numer, and the date.

Example:

Ms. Jane Folley   –2–   July 10, 1986

Complimentary Closing

Locate the complimentary closing two spaces below the body of the letter. Although many options are available, the most common expressions are *Sincerely*, *Sincerely yours*, *Yours very truly*, and *Very truly yours*. Capitalize only the first word and follow the expression with a comma.

Signature

Sign your name four to six spaces below the complimentary closing. Type your name immediately below your signature.

If your name is not clearly male or female, you have the option to include a title of respect—i.e., Ms., Miss, Mrs., or Mr.—in parentheses to the left of your typewritten signature. This information may help someone to address you properly in return mail.

The name of your company may appear in the signature. If you place it first, responsibility for the letter lies with the company. If you place your name first, responsibility lies with you.

Example: Sincerely yours,
Dominion Drilling Company

or

Oscar C. Calloway
(Name of company here)

Optional Data Lines

Three common variations are (1) Identification, (2) Enclosure, and (3) Copy lines.

IDENTIFICATION LINE:
If you do not type your own letter, have your typist include an identification line located two spaces below your signature, at the left margin. Included are

your own initials (sender) in capitals, followed by a colon, and the initials of your typist in lower case.

    Example: WDC:mp

ENCLOSURE:

When you enclose something with your letter, type an enclosure line two spaces below the identification line, at the left margin. If you do not have an identification line, place the enclosure line in its position two spaces below the signature, at the left margin. Some writers simply write *Enclosure;* others follow with a colon and list what is enclosed.

    Example: Enclosure

                 or

                 Encl.: April report

                 or

                 Encl. (3)

                 or

                 Enclosures: April and May reports, Program
                               Evaluation, and June/July
                               projections

COPY LINE:

When you send a copy of your letter to another person, for the sake of courtesy include a single or double letter *c* (lower case) at the left margin, followed by a colon. Place the copy line two spaces below the previous entry. After the colon, type the name or names of those receiving copies.

    Example: cc: J. P. Jones
                       M. D. Weisburg

TONE AND WORD CHOICE

Business letters substitute for a personal conversation between you and the person to whom you are writing. Because of the limited audience and the nature of the document, letters are generally more informal and personal than other business and technical documents. Business letters are characterized by a *you* orientation, positive tone, and natural language.

You Orientation

In a letter, use the pronouns *you* and *I* to obtain a personal, conversational quality, sometimes called a *you orientation*. Some authorities suggest a ratio of approximately three *you's* for every *I* to avoid sounding self-centered and to show your interest in the reader. Actually, if you keep the ratio about one to one, your writing will sound conversational and personal. The damage caused by failing to maintain a *you orientation* is illustrated by Figure 10.3.

```
To Whom It May Concern:

     I am seeking employment as a draftsman. I was raised in a
small rural community and have a work experience which is pri-
marily related to that background. However, I have also en-
joyed the drafting I have done and would like very much to
fully explore that facet of myself and use my educational
training in a full-time capacity.

     I am interested in any opportunity which may arise in the
next calendar year for achieving this purpose.

     I am looking forward to hearing from you.

Sincerely,

XXXXXXXXXXXXXXX
```

Figure 10.3 Example: Poor I/You Ratio

This letter contains eight *I's* (counting variations—*my* and *myself*) and one *you*. The author seems interested in what his potential employer can do for *him*—not what he can do for the potential employer. This letter fails for a number of reasons: the author does not sell himself, gives no evidence why he should be hired, and uses a poor *I/you* ratio, making the letter sound self-centered.

Positive Tone

Another part of the *you orientation* is writing with tact and courtesy. Your letters should sound positive and friendly in nearly all circumstances—even those for which your good will has worn a bit thin. Compare the following examples:

Original: As you were told in our last letter, this office has no jurisdiction over claims.

—The tone is rather like that of an impatient parent speaking to a child. This is a sure-fire method to antagonize customers!

Revision: I'm sorry, but our office doesn't handle claims. Please write the Claims Department at this address. . . . I'm sure they can help you.

—The tone in the revision is friendly and helpful.

Original: In your letter of January 19 you agreed to consult with our Collection Department, but you neglected to set the date you were available.

—Just one word—*neglected*—produces a negative tone.

Revision: We're happy that you have agreed to consult with our Collection Department. Please let us know when you will be available.

—The revision has a positive approach to the same situation.

If you want your readers to respond favorably to your letters, avoid negative words like these:

blame	dispute	impossible
cannot	disagree	neglect
complaint	fail	never
contend	fraud	unable

Rather, use positive words such as these:

can	prompt	thanks
credit	recommend	truth
gain	service	valuable
help	success	vital

In strengthening your letters with positive words, you do not have to disguise your meaning. The idea is to approach your subject positively, accenting what is right and good.

Natural Language

A letter should sound natural and conversational. Levels of formality vary according to your analysis of the reader and situation, but you should aim for a friendly, conversational quality and avoid letter clichés such as the following:

Original: This is to acknowledge that your letter of November 5 has arrived at our office.

Revision: Thanks for your letter of November 5.

Here are some letter-writing cliché's (worn out expressions) to avoid:

CLICHÉ	NATURAL/PLAIN ENGLISH
enclosed herewith please find	I have enclosed . .
I have before me	(Cut as deadwood)
I wish to state	I believe . . .
in response to the same	In reply, I . . .
In view of the fact that	Because . . .
in the present writing	Now . . .
Please be advised that your payment is now due.	Your payment is due.
As per your request	As you requested . . .
This writer	I
I am cognizant of the fact that . . .	I know that . . .
in the immediate future	soon
due to the fact that . . .	because . .
Herein enclosed	Enclosed

ORGANIZATION

Writing a business letter is similar to writing most other documents. A letter has an introduction, body, and conclusion; it emphasizes key ideas in the introduction and conclusion; and it usually includes the subject, purpose, scope, and plan of development in the introduction. A letter differs by being more conversational and personal.

This section explains how to write a business letter by analyzing the audience, anticipating the reader's questions, and providing answers for them within the traditional three-part organization.

Audience Analysis

Before writing any document, you should analyze your audience so you can choose appropriate language, subject matter, and details to meet the readers' needs. Because a letter is usually written to just one person, the opportunity and need for careful audience analysis are greater than for most documents.

Before writing a letter, ask yourself these basic questions:

1. Who is my reader?
2. What is the reader's job?
3. What is the reader's title?
4. What does the reader know about the subject?
5. What is the reader's attitude toward the subject?

Answering these questions will help you meet the needs of the reader and, at the same time, achieve your goal of clear communication.

The following paragraphs explain how analyzing and anticipating your reader's questions and needs will help you to write the introduction, body, and conclusion to a business letter.

Introduction

The first paragraph of a letter is the introduction. Because it is in an emphatic position, it contains important information. At the beginning of a letter, the reader needs answers to the following questions:

Introduction: —Who are you?
            —What do you want?
            —Why are you writing me?

Study this example:

Example: ACME Construction, as part of a larger proposal, is seeking bids for the design, supply, and erection of a large metal building as part of an air separation plant to be constructed in XXX, Iowa.

Analysis: —WHO IS WRITING? The letter head and signature clarify this point.
            WHAT DO YOU WANT? ". . . design, supply, and erection of a large metal building"
            —WHY ARE YOU WRITING ME? ". . . seeking bids"

In terms of the four standard elements of an introduction, the example breaks down this way:

1. *Subject*—"design, supply, erection of large metal building"
2. *Purpose*—"seeking bids"
3. *Scope*—The letter does not specifically state *scope*. This is not unusual in such a short document, especially when the content of the body is so clearly implied—it involves specifications of the building.
4. *Plan of Development*—This letter does not state a plan of development. This is often true of short documents.

Always include minimally the *subject and purpose* in the introduction to your letters. This will prepare the reader for the detail you include in the body.

Body

What you state in the body of a business letter depends upon what you promised in the introduction. In terms of meeting the reader's needs, attempt to answer questions such as these:

Body: —What are the key points, parts, aspects of the subject?
        —What are some specific details of
                cost
                steps
                reasons
                dimensions
                location (where)
                schedule (when)
        —How does it work?

Now study the relationship between the introduction and the body of the sample letter:

Example: The building is 360 feet long by 110 feet wide by 68 feet high. A 60 ton overhead crane will be installed at a height of 60 feet and run the entire length of the building. I am enclosing a description of site conditions and a foundation print. As more information becomes available I'll forward it to you, although specifications indicate the bidder is responsible for design.

The body contains what the introduction promised—the answers to the reader's questions about building dimensions, facilities, and so on.

Conclusion

The conclusion of a letter may be a single sentence signaling to the reader that you have reached the end of your message or, more likely, several sentences emphasizing important information from the introduction and/or body. Again, answering the reader's questions is the key to complete communication:

Conclusion: —What is most important for me to remember from the discussion?
                —Is there a deadline or key requirement I should remember?
                —Is there a specific response you want from me?

Study the conclusion from the sample letter to see how it meets the needs of the reader:

Example: Because we have a tight bid schedule, I would appreciate receiving your proposal not later than June 1, 1987. If you have questions, call 208-356-XXXX.

This conclusion emphasizes what is most important to both reader and writer— the due date. By placing such information last, the author helps the reader recognize and remember the information. The last line, listing the telephone number, is also functional because the tight bid schedule requires quick communication.

TYPES OF LETTERS

Business and technology use many types of letters, far more than can be discussed in this section. However, no matter what the need, if you analyze your audience and the communication situation, and use three-part organization to meet the reader's needs, you will communicate successfully.

This section briefly discusses writing letters of inquiry, order, and adjustment; and concludes with a detailed discussion of the letter of application and résumé.

Letter of Inquiry

When you need information—for example, a brochure, catalogue, price, data on availability, and so on—a letter of inquiry sent to an individual or company will usually obtain it for you. Here are a few basic guidelines.

Letter of Inquiry

1. Use three-part organization.
2. State subject and purpose in the first paragraph.
3. Be specific about the information you want. State model or catalogue number, etc.
4. State how providing the information may benefit the company or individual answering the request.
5. Thank the reader for helping you.
6. Include a self-addressed stamped envelope when writing to an individual.

Study Figure 10.4.

Order Letter

As an individual or as a representative of a company, you may need to order or purchase something by mail. An effective order letter provides accurate information about what you want, the price, how you will pay for it, and how you want it delivered. Here are some basic guidelines:

P.O. Box 789
Rexburg, ID 83440
February 3, 1985

Cuesta Systems, Inc.
5440 Roberto Court
San Luis Obisop, CA 93401

The December 1984 PC World contains an advertisement for Data-
saver, a backup power unit for computers. I am interested in
using such a device on my IBM–PC.

Please send information on size, weight, power capability,
operating duration, alarm systems, and cost.

Because I live in an area subject to frequent power outages,
your product may be just what I need to protect my data files.

Thank you.

Mary Folley
Mary Folley

Figure 10.4 Letter of Inquiry: Simplified Block Format

Order Letter

1. Use three-part organization.
2. Clearly state what you want. If ordering two or more items, use an itemized list that includes catalogue and model numbers, size, color, weight, and so on.
3. Explain the price you expect to pay and how you'll pay it: check, credit card, COD, charge your account. For a credit card the seller needs your name, account number, and expiration date.
4. Explain how you want the item shipped: air express, parcel post, truck, or other means.

Study Figure 10.5.

Claim and Adjustment Letters

If all orders were filled on time and exactly as requested, and if all customers paid the full amount promptly, claims and adjustment letters would indeed be rare. However, misunderstandings do arise, shipments are sometimes late, payments come overdue, and parts or equipment must be back-ordered. When these problems arise, it is wise to write calm, tactful, courteous claims and adjustment letters. Here are basic guidelines for writing them:

ACME   2550 North Sixth Street
                            Marblehead, MA 01945

                            September 23, 1986

Edward T. Enfield
Circulation Director
IFM
IFM Building
Old Saybrook, CT 06475

Dear Mr. Enfield:

    Please send one copy of your company's publication, The
IFM Guide to Preparation of a Company Policy Manual.

    Because we are in the midst of revising our own policy man-
ual, we are anxious to receive the information contained in
your publication. Therefore, I am enclosing a check for $94.95
to cover both the purchase price ($64.95) and charges for air
express shipment.

    Thank you.

                            Sincerely,

                            Mary K. Messer

                            Mary K. Messer
                            Personnel Director

MKM:jd

Enclosure: Check $94.95

Figure 10.5   Order Letter: Modified Block Format

Claim Letter

1. Check catalogue or other sources for explanation of return and claim procedures. Many firms have a standard claim form that you should use if at all
   possible.
2. If you must write a letter, use three-part organization.
3. Be fair and reasonable. Focus on the issue.
4. Clearly identify the transaction by including purchase order number, copies
   of invoices, canceled check, or other documentation.
5. State the adjustment or action you feel will solve the problem.

Adjustment Letter

1. Use courtesy and tact in answering a claim letter.
2. Show your concern by answering promptly.

3. Clearly identify the claim letter and specific transaction involved.
4. Explain what action you are taking. If you cannot honor the claim, explain why.

Figure 10.6 is a claim letter, and Figure 10.7 is the answering adjustment letter.

Letter of Application and Résumé

Sooner or later you will need to apply for a job. One common way is to send a letter accompanied by a summary of your qualifications. This summary is called a *data sheet*, *vita*, *dossier*, or *résumé*. Because these documents may be your first contact with a potential employer, they must be neat, specific, and well

---

Route 1, Box 98
Craig, CO 81625
January 28, 1986

Outdoor Fashions
Catalogue Department
P.O. Box 4333
Chicago, IL 60647

Outdoor Fashions:

Two weeks ago I ordered the two-piece suit featured in your winter sale catalogue. I am returning it for replacement or refund because serious faults in construction make it unwearable.

The main seam in the back of the coat is crooked, probably because the material slipped when it was being sewed.

My original order was for items M-1230 (coat) and M-1231 (skirt) in medium blue at the sale price of $134.95 (paid by check with the order). I would still like the suit if it is available. If not, please fully refund the purchase price.

Thank you.

Sincerely,

Sue Brown
Sue Brown

Figure 10.6   Claim Letter: Block Format

February 7, 1986

Ms. Sue Brown
Route 1, Box 98
Craig, CO 81625

Dear Ms. Brown:

We're sorry to learn of the problem you've experienced with
the suit ordered from our sale catalogue. We want you to be
satisfied with your order.

Due to the popularity of the two-piece suit, it is temporarily
out of stock for at least three weeks. Therefore, we are fully
refunding the purchase price of $134.95.

If you should care to resubmit your order, we'll be happy to
send you a replacement at the sale price as soon as new stock
arrives.

We apologize for the inconvenience this may have caused you.
Thank you for shopping at Outdoor Fashions.

Yours very truly,
OUTDOOR FASHIONS

Mary B. Beck
Mary B. Beck
Customer Liaison

MBB:kc

Encl.: Check for $134.95

Figure 10.7   Adjustment Letter: Full Block Format

written. They should present your qualifications in the best possible light; after
all, the only reason you write them is to sell yourself to a potential employer.

The first part of this section discusses key features of the letter of application—
audience analysis and basic organization; the second part discusses the four
basic parts of the résumé.

AUDIENCE ANALYSIS

Before attempting to write a letter of application, learn all you can about your
potential employer. What product or service does the firm sell? What positions
are available? What skills and training do employees need? What are the working
conditions? What is the starting pay scale? What is the future of the company—
and its employees? Who hires?

You need to gather enough information to make a realistic decision about whether you want to seek employment with a specific firm. You need to know whether you skills and training give you any chance of getting a job there. Here are some simple ways to gather information:

1. Talk to present and former employees. Find out what demands and skills the job requires.
2. Talk to others employed in a similar field.
3. Visit a public or college library to research a company and specific career field. Examine books such as the following:

   —*The Encyclopedia of Careers and Vocational Guidance* (2 vols.) edited by William E. Hopke.
   —*Occupational Outlook Handbook*, a yearly U.S. Department of Labor bulletin.

As you do your research, make a list of the skills and training required for employment in a specific job or field. Once this list is complete, make a separate listing of your own training and experience; list every job you have ever had; list your education—including courses that may be relevant; list specific skills, special interests—anything that may link you to the job you are seeking.

Now compare the two lists and circle any match-ups between job requirements and your qualifications. These are key points to stress in the letter of application and in the résumé. If you have no direct experience, what are your best qualifications? If you are just starting in a particular field, your college training (specific courses and skills) may be the best information to stress for this audience.

Knowing what your audience is looking for, and tying yourself to those interests—that is what writing a letter of application is all about.

ORGANIZATION

A letter of application, similar to most business letters, uses a standard format—block, semi-block, modified block; three-part organization; and high standards of typing and layout. Here are some basic guidelines:

Introduction

1. Make the first paragraph short—no more than three sentences.
2. Explain how you heard about the job—newspaper, employee, friend, relative, employment agency, college financial aid office, and so on.
3. State the job or position you are seeking.
4. Make explicit application for that job. See the first paragraphs of Figures 10.8 and 10.9.

Body

1. Tie yourself to the position for which you are applying. Distinguish yourself from all other applicants by detailing one or two qualifications (your best:) that make you a good candidate for the job. Do not try to tell everything

about yourself—save this information for the résumé. Study the second paragraphs in Figures 10.8 and 10.9.

Conclusion

1. Close courteously.
2. Ask for an interview and really sound like you want one. Remember, an interview is about all you can realistically hope to gain from your letter.
3. Make it easy for an employer to contact you. Give your telephone number and tell when you will be available. Study the last paragraphs of Figures 10.8 and 10.9.

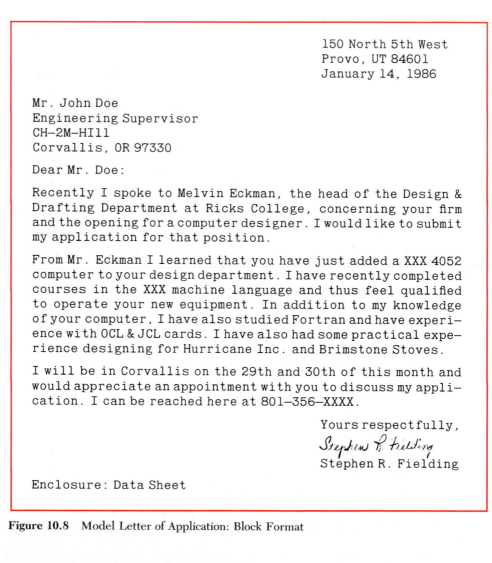

```
                                        150 North 5th West
                                        Provo, UT 84601
                                        January 14, 1986

        Mr. John Doe
        Engineering Supervisor
        CH-2M-HIll
        Corvallis, OR 97330

        Dear Mr. Doe:

        Recently I spoke to Melvin Eckman, the head of the Design &
        Drafting Department at Ricks College, concerning your firm
        and the opening for a computer designer. I would like to submit
        my application for that position.

        From Mr. Eckman I learned that you have just added a XXX 4052
        computer to your design department. I have recently completed
        courses in the XXX machine language and thus feel qualified
        to operate your new equipment. In addition to my knowledge
        of your computer, I have also studied Fortran and have experi-
        ence with OCL & JCL cards. I have also had some practical expe-
        rience designing for Hurricane Inc. and Brimstone Stoves.

        I will be in Corvallis on the 29th and 30th of this month and
        would appreciate an appointment with you to discuss my appli-
        cation. I can be reached here at 801-356-XXXX.

                                        Yours respectfully,

                                        Stephen R. Fielding
                                        Stephen R. Fielding

        Enclosure: Data Sheet
```

Figure 10.8   Model Letter of Application: Block Format

```
                                        Ricks College
                                        Dorm 3, Apt. 93
                                        Rexburg, ID 83440
                                        October 15, 1986

Mr. Henry Doe
Farmland Industries
P.O. Box 7300
Kansas City, MO 64116

Dear Mr. Doe:

    In a recent conversation with one of your employees, Mrs.
Cherlyn Jones, I learned of the opening for Technical Secre-
tary in your department. I would like to apply for that posi-
tion.

    Your department is well known for its organization, effi-
ciency, accuracy, and friendliness. I feel I can maintain
these standards. As a student I have taken the required math,
English, and science courses necessary for technical work,
including calculus, statistics, and English composition. I
am also well acquainted with the duties and responsibilities
of a secretary. I have worked as a secretary the past four
summers with many different firms through temporary agencies
and adjusted quickly to the way each office ran.

    I would appreciate an interview to tell you more about my
training and experience. I will arrive in Kansas City on De-
cember 21, 1986, and will be available to meet with you at
your earliest convenience. My phone number here in Idaho is
208-356-1415.

                                   Sincerely yours,

                                   Jeanne C. Monroe
                                   Jeanie C. Monroe

Enc.: Résumé
```

Figure 10.9 Model Letter of Application: Modified Block Format

Résumé

A résumé is a one- or two-page summary of important information about you personally and about your experience, education, and references. The letter of application introduces you to the prospective employer, and the résumé fills in details about your overall qualifications.

Again, neat appearance, accurate punctuation, correct spelling, adequate margins, and general readability are important. Avoid cramming so much data on the page that the reader has to search to find specific information. A good solution to the readability problem is to use major headings and several levels of indentations. White space and careful organization of material also help. Your goal should be to produce a document that can be read quickly and easily.

There is no single standard format universally used for writing résumés. Depending upon audience analysis and previous experience, some writers create a document that they feel represents them best. This makes good sense because the résumé, like the letter of application, is a sales document, selling *you!*

This section discusses a simple (but sound) basic résumé format that includes heading, position desired, education, experience, personal data, and references. Here are some basic guidelines:

Heading

1. At the top of the page state your full name, mailing address, and telephone number (including area code), and the date. Center data or place at the left margin.

Position Desired

2. Identify the position you are seeking.
   Position Sought: Computer designer

   The strategy here is to identify yourself and immediately relate yourself to a particular job. If you are not applying for a specific position, delete this line.

Education

3. In reverse chronological order, list all your education, including high school. For college, list the name of the school, when you graduated (or expect to), degrees earned (or anticipated), major and minor fields, and any courses that may particularly relate to the job applied for. For high school, name the school, location, and date you graduated.
4. Stress your accomplishments. If you have a good grade point average, tell about it. If it is poor, do not mention it. If you have earned honors or awards, mention them. Be honest in what you write; but stress your strengths, not your weaknesses. Based on your analysis of the employer's needs, you may want to give some detail about one of your college research projects if it will help tie you to a job.

Experience

5. Use reverse chronology to list all your work experience. If the list is extensive, delete some of your first jobs and concentrate on recent experience. For each position, list the following:

— dates of employment
— position
— type of work
— name of company
— name of supervisor
— address

If you list some of your previous employers as references, delete the address in this entry to save space and avoid needless repetition.

6. If one or several of your previous jobs particularly relate to the position you are applying for, stress them by pulling them out of the reverse chronological order and giving them a heading of their own. The heading might be *Related Experience*, placed just before the section on experience. This is part of selling yourself—helping readers to see how you relate to their job openings.

References

7. List three references. Choose those who will support you—ignore your enemies! Traditionally, references have included one each for work, education, and character. Practice varies widely. Analyze your situation, and use what fits your needs. For each reference, include the following:

— full name
— title (if any—i.e., "Personnel Director")
— address (include zip code)
— area code and phone number

Before listing people as references, contact them and ask permission to use their names. At the same time, take a few minutes to tell them about the job you are seeking and why you want it.

Vary the order in which you present education and experience in the résumé, depending on which best qualifies you for the job. If you have strong related work experience, put it before education. If not, stress education by putting it first.

As a college student, you may not have any directly related work experience. If this is the case, list *any* job you might have had. Many employers may be interested in the fact that you have held a steady job of any kind. If you have little work experience, stress your education by going into some detail of related course work and your job goals. There is no excuse for a skimpy letter of application or résumé.

Study the model résumé or data sheet presented in Figure 10.10.

```
                          DATA SHEET

                        Stephen R. Jones
                       150 N. 5 W., Apt. 87
                        Rexburg, ID 83440
                     Telephone 208-356-XXXX
                          January 1986

POSITION DESIRED:      Computer designer

EDUCATION

     Ricks College, Rexburg, ID
        Graduation: April 1986
        Grade Point Average: 3.8 of 4.0 (Dean's List)
        Major: Design & Drafting Technology

     Important Courses
        Adv. Engineering Graphics, Technical Illustration,
        Technical Mathematics, Computer Graphics, Fortran
        Technical Writing

     Snake River High School, Blackfoot, ID
        Graduation: 1982
        Grade Point Average: 3.6 of 4.0 (Honor Roll)

WORK RELATED EXPERIENCE

     1985 Part-time designer for Brimstone Stoves
          Wayne Gunderson, President
          Mannan, ID 83442
          208-352-XXXX

     1984-1982 Part-time designer for Hurricane, Inc.
          Dean Kirtpatrick, President
          Blackfoot, ID 83221 (See References)

PERSONAL INFORMATION:

     Age: 23              Hobbies: skiing, baseball,
     Height: 5'11"                 racquetball, hunting
     Weight: 155 lbs.              and fishing
     Married: no children

REFERENCES

Mr. Melvin Eckman     Mr. Dean Kirtpatrick   Mr. Blair Pincock
Drafting Dept.        Hurricane, Inc.        Drafting Dept.
  Coor.               523 S. 300 W.          Ricks College
Ricks College         Blackfoot, ID 83221    Rexburg, ID 83440
Rexburg, ID 83440     208-624-XXXX           208-356-XXXX
208-356-XXXX
```

Figure 10.10   Model Data Sheet

REVIEW QUESTIONS _____

1. In what way can a letter be considered a public relations document?
2. What are the four common formats for business letters?
3. How do the four common formats differ?
4. What are the minimal margins for a business letter?
5. What is the typical pattern of spacing within a letter?
6. What information should you include in the inside address?
7. When is it appropriate to use an attention line?
8. What are several ways to avoid sexist language in the salutation of a letter?
9. What is the proper way to write a heading for the second page of a letter?
10. Which words are capitalized in the complimentary closing?
11. In the complimentary closing, how can you show individual and company responsibility for a letter?
12. What do the initials on the identification line (end of a letter) mean?
13. What is an appropriate tone for a business letter?
14. What is a *you orientation?*
15. What is a desirable *I/you* ratio in a business letter?
16. How is the organization of a letter similar to that of other business documents?
17. What is the principal difference between a letter and other documents?
18. Why is audience analysis particularly important in a letter?
19. What is typical content for the first paragraph of a letter?
20. What is typical content for the body of a letter?
21. What is typical content for the conclusion of a letter?
22. When is a letter of inquiry useful?
23. When is an order letter useful?
24. What is the appropriate tone for claim and adjustment letters?
25. What is the basic purpose of audience analysis before you write a letter of application and résumé?
26. What is the typical content of the first paragraph in a letter of application?
27. What is typical content for the body of a letter of application?
28. What is the typical content for the conclusion for a letter of application?
29. What are the basic parts of a résumé?
30. How does the purpose of the résumé differ from that of the letter of application?

EXERCISES

1. Write a letter to a former employer asking permission to use the person's name as a reference. Provide some basic information about the position you seek and your interest in it. Use a standard letter format.

2. Write an order letter for something you have seen advertised in a magazine. Use a standard letter format and three-part organization. Follow the guidelines for writing an order letter.

3. Write a claim letter. Assume that a product you have owned less than a month is faulty (i.e., portable radio, hair dryer, etc.) in some way and you are seeking a replacement or refund. Use a standard format and three-part organization. Follow the guidelines for writing a claim letter.

4. Analyze the following letter for tone. Which lines would be particularly objectionable to the addressee? What changes would make the tone more friendly but still firm?

    Dear Mr. Corrigan:

    We have received your letter regarding J. J. Jefferies as a candidate for Assistant Process Facilities Manager, San Francisco.

    Jeffries seems a good solid engineer with the proper mix of design and project responsibility. However, as you note, he does not have oil and gas experience. Further, he does not have petrochemical experience, and his synfuels experience is miniscule. You should have understood he is unsuited to participation in ACME Hydrocarbon Processing as a leader. And, as was explained to you at the time of review of your draft of the position specifications for San Francisco, such suitability is a prerequisite. Hydrometallurgical experience has some application, but the methodologies of hydrocarbon processing are so different that the base experience in hydrocarbons must be there.

    Jeffries is not suitable for any of the positions we have defined.

    We are also using this opportunity to inform you that we have been able to fill the position of Assistant Process Facilities Manager from another source. Please consider that requisition now cancelled.

    Very truly yours,

5. Analyze the following sentences for tone. Underline any words that are too informal, hostile, ingratiating, blunt, sarcastic, whining, evasive, negative, and so on. Now rewrite the sentences to produce an acceptable tone.
    a. Although questionnaires use valuable time, I hope you won't throw this one away.
    b. I realize your job is demanding, but I hope I can convince you to join one more committee.
    c. You state your order arrived two months late.
    d. You contend that the equipment we sent you is totally inadequate for the job.

    e. As I told you in the last letter, our firm was not responsible for the design of that project.

    f. I have received your complaint about our delivery policy which you claim is negligent and inadequate.

6. Using the models and suggested formats and guidelines, write a letter of application for a job. Make the letter perfect in spelling, grammar, typing, and format. Apply for a summer job or a full-time position. Be as realistic as possible—ideally, write a document you actually plan to send.

Use accurate information about your present education and experience to apply for a job within your range of competence.

Perform enough research on a potential employer to be able to tailor your letter to highlight your best qualifications for the job.

7. Write a résumé to accompany Writing Assignment 6. Use the models and guidelines to prepare a document sufficiently accurate in detail and pleasant in appearance to be useful in applying for a job. Use accurate information about your present experience and education.

11

Memos

A memo, or *memorandum,* generally used for communication within a firm, is the chief means of sharing data, making recommendations and requests, clarifying progress, and maintaining a written record. Memos are similar to letters in tone, language, and organization, but they differ in format and use.

Documenting or creating a written record is one of the main functions of memos. Company policies, guidelines, and procedures frequently need clarification and revision to meet changing needs. This information, shared in the form of a memo, creates a written record. Later, if needed, the author or reader can pull the memo from a file to substantiate what was said.

This chapter discusses basic memo layout, common parts, organization, types of memos, and concludes with review questions, and revision and writing assignments.

LAYOUT

Like all business and technical documents, memos need attractive spacing, margins, and adequate white space. Because most firms use a standard memo form, it usually is not necessary to create your own. Although layout varies in detail with different companies, Figure 11.1 clarifies basic elements of spacing, location, and margin.

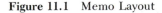

```
                          COMPANY NAME
                          MEMORANDUM


        DATE:      May 1, 1986
        TO:        Name,
        FROM:      Name, (initials optional)
        SUBJECT:   Write a clear title in capital letters.

        Introduction—single spaced
        Body—single spaced
        Conclusion—single spaced
```

Figure 11.1   Memo Layout

PARTS

As shown in Figure 11.1, the memo has only two parts: the heading material and the body. Although a memo has no salutation or complimentary closing (as does a letter), it is similar in sometimes having a signature and optional lines to identify writer and typist, enclosure, and copy distribution.

Heading

1. Give the *date*, *to* and *from*, and the *subject* (preferably in an order that places the subject just before the body).
2. Write the subject explanation in capital letters. Use carefully selected words to identify the subject of the memo clearly.

Body

3. Use three-part organization: introduction, body, and conclusion.
4. Use the same format as a letter to direct the reader's attention to another page.

   Example:

   J. E. Brown   –2–   2/15/86
   or
   J. E. Brown
   2/15/86
   Page 2

Optional Lines

5. *Signature:* Although most writers do not sign a memo after the body of the document, practice varies widely. For example, an administrator establishing policy may sign to show specific responsibility for the content. Many writers prefer to pen their initials next to the typed *from* entry at the top of the memo. The signature, when used, is located two to four spaces below the body.
6. *Identification line:* As with letter usage, this optional line appears two lines below the signature.

   Example: WDC:lr

   The capital letters are the initials of the author; the lower case are the initials of the typist.
7. *Attachment* (Enclosure): As with letter usage, this optional line appears two lines below the identification line. Typical form is *Attachment:* followed by a colon and titles of documents attached. The items are usually stapled to the memo.
8. Copies: Placed two spaces below the last entry, a *copy* or *distribution* line clearly identifies who should receive a copy of the memo. At a glance, the reader can tell who else shares in the information. Common forms are as follows:

   Example: c: J. H. Brown
             R. W. Groom
     or
     cc: J. H. Brown
             R. W. Groom
     or
     Distribution: J. H. Brown
                R. W. Groom

Many times memos circulate to hundreds of people within a company. In this case, a code is useful (see the *To* line in Figure 11.2). For example, a distribution list may use the letters A,B,C,D. Memos marked *A* on the copy or distribution line go automatically to all employees; *B* restricts distribution to supervisors; and so on.

Study Figure 11.2 for an example illustrating the use and position of common memo parts.

ACME INTEROFFICE CORRESPONDENCE

Date:	September 29, 1986
To:	Distribution PA, PB, PC and PD
From:	Randy Thompson
Subject:	PROBATIONARY PERIODS

In accepting employment with ACME OIL all employees are subject to two probationary periods. I would like to clarify this policy further.

The first 180 days of employment for all permanent employees (both new hires and re-hires) is a probationary period during which the company evaluates the employee's ability to perform the work assigned. This probationary period relates only to job performance.

In addition, all permanent employees (both new hires and re-hires) serve a concurrent 90 day probationary period for attendance only.

If you have any questions, please call extension 223.

Randy Thompson

Randy Thompson
Personnel Manager
RT:np
cc: Ron Thomason

Figure 11.2 Sample Business Memorandom

ORGANIZATION

The organization of memos is similar to letters and other documents; that is, they all use three-part organization. Like the letter, memos are often written to individuals and, therefore, tend to be personal. However, practice varies widely, based primarily upon purpose. For example, an engineer writing a memo proposal, which will circulate widely in management circles, should use care to select a level of formality appropriate for the situation and intended audience.

This section discusses audience analysis, basic organization, and ways to improve readability of memos.

Audience Analysis

Audience analysis is just as necessary for memos as it is for other documents. Before writing a memo, ask yourself these basic questions:

1. Who is my reader?
2. What is the reader's job?
3. What is the reader's title?
4. What does the reader know about the subject?
5. What is the reader's attitude toward the subject?

Unlike most business letters, which are addressed to individuals, memos sometimes circulate widely within a company, often well beyond the distribution list. This is an aspect of life in an organization that beginners often overlook.

For example, consider this hypothetical situation:

> You are working as an engineer on a project. Your supervisor, someone you know well, asks you to write a report explaining why some machinery is not working properly. You quickly analyze the situation and scratch out a short note to your supervisor, not paying much attention to form or writing style—after all, your supervisor knows what you are talking about.
>
> The supervisor, needing documentation for a follow-up report, sends your short memo up the chain of command. Because the original incident was important to production, your memo may eventually reach the president of the company. If you had realized this, you would have been much more careful in writing the original.

Because everything you write represents you—in the classroom and on the job—consider both the primary audience and any secondary audience. Attempt to meet the needs, background, and questions of both audiences.

```
                        NEW EMPLOYEE
Skip,

    Paul Lander will be in at 8:00 Friday. He will need a stain-
less steel Riley Power Bin and 5 tons of 16-20-0 fertilizer.

    Check the oil, gas, tarp, spout, and start the engine of
the power bin this afternoon to be sure it's working properly.

    This afternoon check the overhead bin to be sure 15-20 tons
of 16-20-0 are available. Other customers are sure to need
some also.

Lou



                     EXPERIENCED EMPLOYEE
Skip,

    Paul L. will be in first thing in the morning. He needs a
Power Bin (Stainless Riley) and 5 tons of 16-20-0.
```

Figure 11.3   Simple Memo: Impact of Audience Analysis

Figure 11.3 illustrates how a writer can adjust to the needs of the reader. The example is a short note or memo addressed to one employee but written in two forms: the first to a new employee, the second to an experienced employee. The author is foreman of a five-man crew at a farm chemical distribution center where informality is appropriate.

Although the example is extremely simple, it teaches an important lesson. The *new employee* version assumes little knowledge: it clarifies the customer's name, the time, the equipment needed, and what actions to perform before the customer arrives. This version meets the needs of a particular reader. The *experienced employee* version is brief because the reader knows what to do.

As a writer, always attempt to meet the needs of your audience. Careful analysis will help you communicate more effectively.

Basic Organization

The organization of a memo is similar to that of other business documents. When you think *memo organization*, think *three-part structure*. (For a review, see Chapters 2 and 3. Even if a memo is just three sentences long, try to make the first an introduction, the second the body, and the third the closing or conclusion.

The guidelines for writing letters discussed in Chapter 10 are equally effective for memos. Your basic strategy in writing should be to meet the needs of the reader by supplying the information needed for clear communication. Here are some questions that readers may ask. Obviously not all of them are appropriate for a given memo:

Introduction

- Why are you writing me? (Purpose)
- What is this memo about? (Subject)
- What aspect of the subject are you focusing upon? (Scope)
- How have you divided or classified the ideas you are telling me about? (Plan of Organization)
- Can you give me an overall idea of where you are going with this subject?
- Are there conclusions or key findings you can describe?

Body

- Based upon what you have told me in the introduction, I would like to know some specific details about
  - cost
  - approach
  - steps
  - reasons
  - dimensions
  - location
  - schedule
  - what it is
  - how it works
  - why you did it that way
  - who was involved
  - justification

Conclusion

- Could you pull it all together for me again before you stop writing?
- What are the conclusions?
- What are your recommendations?
- What should I remember?
- Is there a deadline I need to meet?
- What should I do now that I have read your memo?

By giving a memo three-part structure, you meet the reader's need for organization and structure. Failure to do so often causes misreading and frustra-

Date: April 8, 1986

To: Bill Masters

From: Phil Dean, Engineer Regulatory Compliance

Subject: MHA Expansion—1986

Carl Ranstrom asked me to relay to you in his absence the attached permit application exhibit and following related information regarding the subject expansion. There are three principal areas of expansion, justified as follows:

1. MATERIAL LAYDOWN PAD—A 1000' × 1300' pad and associated access roads are proposed at the west end of the existing MHA pad. The pad is to be used by the construction group for material control and storage, and is required due to several factors:

   —dramatic increase in construction activity
   —more storage needed
   —loss of storage areas at PS 12 and PS 13

   The pad will increase operating efficiency of material handling due to consolidation of storage and central location to the field.

2. H-1 PAD—An expansion of the existing MHA pad is proposed on the north side, accomplished by laying gravel between two existing gravel pad arms. The new pad will accommodate proposed additional storage and warehouse needs.

3. LIVING QUARTERS—Expansion of the existing pad is proposed by laying gravel to the west of the existing living quarters. This expansion is required to accommodate 4 additional sleeper wings to the existing quarters.

Start—up for the project is slated for June 1, 1986.

Please forward these exhibits along with a permit application for the project as replacement for the previously submitted rectangle application for the MHA, emphasizing the need for permits by June 1, 1986.

This answers why

5/20

Figure 11.4   Sample: Effective Memo

tion. Figure 11.4 is an example of an effective memo that uses three-part organization and, at the same time, that answers key questions for the reader. The result is clear communication.

The first paragraph of Figure 11.4 is the introduction, the three numbered sections constitute the body, and the last four lines are the conclusion. The introduction clarifies the subject, purpose, and scope. The body provides detailed information and specific support for each aspect of the expansion. The conclusion tells the reader what to do with the information and attachments, stressing the importance of the June 1, 1986 date.

Of course, the writer assumes the reader has some knowledge of the subject involved. For example *MHA*—Main Housing Area—is so well known within the company that it would waste time and space to write it out. The author also uses another short-cut in the first sentence by referring to the *subject expansion*. Some writers, rather than repeat the information contained on the subject line in the heading, simply refer to this data as the *subject* _____. Usually one additional word makes the reference clear. Such a practice is acceptable as long as it does not confuse the reader.

Improving Readability

Another aspect of memo organization is readability. A memo that is a sea of type, crammed to the top, bottom, and side margins in long paragraphs with little white space discourages the reader. Busy colleagues and managers—and teachers—may simply file such a document in the wastebasket as unreadable.

The solution to the problem is to use short paragraphs, several levels of indentation, lists, headings, and white space. Figure 11.4 has an attractive appearance. Imagine how unreadable it would be if the body were all one long paragraph!

Here are a few simple guidelines to help you make your documents readable:

1. Avoid long paragraphs (more than 6–8 lines) by breaking ideas into smaller units.
2. Use lists and numbers to separate items.
3. Use headings to label clearly particular parts of the idea discussed.
4. Create white space with double spacing between paragraphs and several levels of indentation to separate main ideas from supporting ideas.

MEMOS—APPLICATIONS

As stated in the introduction to this chapter, memos are the chief means of communication within a company. In this role they are used for hundreds of purposes, ranging from one-sentence announcements to 8–10-page memo reports. Audience analysis, three-part organization, anticipation of the readers' needs and questions, and attention to general readability apply to all these documents.

The purpose of this section is to provide two additional representative examples of effective memos written for industry. The first, Figure 11.5, is a revision of company policy dealing with recognition parties.

Date: July 2, 1986

To: All Employees

From: Randy Thompson, Personnel Manager

Subject: GUIDELINES—PIZZA RECOGNITION PARTIES

Pizza parties are a way of recognizing employees for their individual efforts to accomplish department goals. These parties should be scheduled on a quarterly basis, when specific goals are met or exceeded.

The following guidelines apply to all Pizza Parties, effective immediately:

1. Parties are set up quarterly.

2. Work shifts will be shortened by one hour only: Days 7 hours, Swing 7 hours, Grave 6 hours.

3. Break times aren't affected.

4. Lunch periods may be skipped if approved by the department manager; however, the hours indicated in "2" above must be worked.

5. Only company employees are invited.

6. All charge slips must be signed by a manager or area supervisor.

If you have any questions, please call extension 1415.

RT/np

cc: J. R. Smith
     M. K. Beck
     S. P. Baker
     P. Fowler

Figure 11.5 Model Memo: Statement of Policy

Figure 11.5 is easy to read because of its three-part organization and effective use of white space and lists.

All examples presented so far in this chapter have been short, consisting of one page or less. Figure 11.6 is a three-page memo dealing with a complex subject. The original document was written by an engineer to an experienced maintenance supervisor. As you review it, do not be discouraged by the technical nature of the subject. Instead, evaluate the organization, basic clarity, readability, use of headings, lists, and so on.

This memo is effective because of its clear organization and its clear statements of goals, objectives, procedures, and responsibilities. The author anticipates most questions readers might ask. The short paragraphs, headings, lists, and open space make the document readable. Because this memo deals primarily with procedure, it does not need a conclusion.

Date: July 1, 1986

Subject: Fin Fan Gearbox Lube Oil Changeout Test Procedures
            and Responsibilities

From: D. B. Thayer, Rotating Equipment Engineer

To: R. L. Foulke

This memo highlights the objectives of testing a different lube oil in several fin fan gearboxes in the MDC and provides procedures for changing the oil and for inspecting the gear-boxes prior to restart. I will also outline the responsibilities of the Maintenance, Operations, and Engineering groups during the test period.

Test Objectives

The objective of this text is to determine whether Chevron Borate 220 gear oil can extend the life of our fin fan gearboxes better than the lubricant now in use, Conoco DN 600.

To perform a comparison we will use Borate 220 in units 18—130 and 18—139, keep DN 600 in units 18—131 and 18—132 for control purposes. The test will last approximately six months to provide summer and winter data.

Cooling bays 120 and 130 were selected for several reasons. The 120 gearboxes have never failed, whereas the 130 units have a long history of failures. Because the two bays are adjacent, this will centralize work involved in the test.

Inspection Procedure

Inspection of the gearboxes should take place with the oil drained, before the addition of the test lubricant. Using

(Continued)

Figure 11.6   Long Memo: Good Example

the normal procedure for inspecting the gearbox in the bay area, measure and record the following:

- Gear wear
- Gear tooth contact pattern
- Motor—to—gearbox alignment
- Fan blade pitch angle
- Bearing condition

Check the bearing by shaking the input and output shafts. Note wobble or noise. Record vibration levels for each gearbox while the unit is running before and after the oil change.

Oil Change Procedure for Units 18—131 and 18—132

Replace the DN 600 in place with fresh DN 600 according to the following procedure:

1. Take a sample of DN 600 in gearbox.

2. Shut down gearbox.

3. Drain old DN 600 from gearbox.

4. Inspect gearbox per above inspection procedure.

5. Fill gearbox with fresh DN 600.

6. Start gearbox.

7. After two hours of run time, sample fresh DN 600 in gearbox.

Oil Change Procedure for Units 18—130 and 18—139

Use a slightly different procedure for putting the Chevron Borate 220 in these gearboxes. The difference is essentially a "flush" cycle to clean out the DN 600 before the Borate is put in continuous use.

To change the oil in 130 and 139, use this procedure:

1. Take sample of DN 600 in gearbox.

2. Shut down gearbox.

3. Drain DN 600 from gearbox.

(Continued)

 4. Add Chevron Borate 220 to gearbox.

 5. Run gearbox for two hours.

 6. Take sample of Borate 220 in gearbox.

 7. Shut down gearbox.

 8. Drain Borate 220 from gearbox.

 9. Inspect gearbox per above inspection procedure.

10. Add Borate 220 to gearbox.

11. After two hours of run time, sample fresh Borate 220 in gearbox.

Organizational Responsibilities

This section defines responsibilities of the Maintenance, Operations, and Engineering groups for this test.

The Maintenance group will perform the gearbox inspection and oil change. In addition, they will mark the boxes containing Borate 220 so the operators will not add DN 600 to them. They will also take samples from each gearbox every two weeks.

The Operations group will maintain the proper oil level in these gearboxes with the correct oil.

The Engineering group will help inspect the gearboxes and record the findings. They will also assist Maintenance and Operations with any problems that may arise. The engineers will interpret the lube oil analysis reports and issue reports on the status of the test.

If there are any questions or comments, please advise.

DBT/ban

cc: (13 names listed on the original)

REVIEW QUESTIONS

1. What are two basic ways memos differ from letters?
2. Are memos usually single or double spaced?
3. What are the four common parts of a memo heading?
4. What is the basic principle for organizing a memo?

5. Do memos have a signature? If so, how is it included?
6. Do memos use identification and attachment lines?
7. How does the audience for a memo sometimes differ from that for a letter?
8. Based upon audience analysis, how would a memo giving instructions to a new employee differ in content from one to an experienced employee?
9. What is the purpose of the introduction to a memo?
10. In relation to the introduction, what is the purpose of the body of a memo?
11. In relation to the introduction and body of a memo, what is the purpose of the conclusion?
12. What are four basic guidelines for improving the readability of a memo?

EXERCISES

1. Study the following short memo:

SAMPLE MEMO

For those departments wishing a brochure for the coming year, would you get in touch with me and arrange a time where we can sit together and plot and plan which may be the best road to take. We are anxious to review and update any brochures which we have. May we hear from you in the near future? We look forward to working with you and know the privilege which is ours in helping adequately cover the water front.

   a. What is inappropriate, even confusing, about mentioning *road to take* in the first sentence and *covering the water front* in the last sentence?
   b. The author has a wordy writing style. Edit the memo by crossing out all unnecessary words and adding information needed to make it complete.

2. The following information, distributed to over a thousand employees, represent the introduction and first few words in the body of a memo on *Emergency Building Evacuation*.

SAMPLE MEMO

Acme's Dallas Safety Committee has been very instrumental in promoting, informing, and training employees in company safety awareness.

In order to refamiliarize yourself with Acme's safety procedures, please review and insure a thorough understanding of the Employee's General Emergency Guide.

You will note that one of the major responsibilities is to educate Acme employees on proper evacuation procedures. Therefore, the following guidelines are provided to assist and possibly save your life during a building evacuation.

- Vacate the building when the alarm sounds.
- Keep calm.
- Shut off all electrical equipment.
- Take personal belongings.

• • •

a. Is the first paragraph an effective introduction to *Emergency Building Evacuation?* Explain.

b. If you were an employee of Acme, would the first paragraph draw your attention and interest you in the memo?

c. Is the second paragraph effective as an introduction to *Emergency Building Evacuation?* Do you think it is likely that many readers of the memo will review their Employee's General Emergency Guide at this time?

d. Which sentence in the first three paragraphs actually introduces the subject of this particular memo? Would a busy reader get this far before throwing the memo away?

e. Imagine you are on the safety committee. Write an effective, attention-getting introduction to this memo.

3. Study the following memo. The subject line is *Job Evaluations Meeting Schedule.*

SAMPLE MEMO

Acme is currently working with Broughton Associates, a consulting firm, to develop a corporate job evaluation program.

Job evaluation is a method of establishing the internal worth of jobs for purposes of salary administration. It is not a performance appraisal of the individual. Based on the competed job evaluations, job descriptions will be written and values will be assigned to each position so that an equitable salary structure results.

You have been selected to complete a job evaluation questionnaire in a meeting scheduled for:

- Day:
- Time:
- Place:

Since you will be expected to fill out the form in the meeting, think about your job—duties, tasks, and responsibilities that you perform on a daily, weekly, monthly or annual basis. Also think about the knowledge and skills that your job requires, key working relationships, and working conditions.

If you cannot attend the scheduled meeting, contact me at Extension 1317.

Revise this memo to give it three-part organization. In the first paragraph (introduction), tell the reader why you are writing and what you want. Provide necessary supporting information in the body.

4. The following document is printed as it was written—all in one long paragraph. The purpose of the memo, according to the subject line, is *Ex-22 Experiment Trouble-Shooting and Repair.*

SAMPLE MEMO

On September 13, 1987, J. Paul and I investigated the cause of the low and erratic counts evident in the daily performance tests on diode No. 2. Substitution of the pileup rejector and main amplifier feeding the multichannel analyzer resulted in no improvement. The original pileup rejector was reinstalled but the replacement main amplifier was retained. An attempt to bypass the analyzer system switch identified that unit as the cause of the problem. The coaxial relay on the rear of the switch module was disassembled and cleaned with alcohol and the contacts burnished very slightly using a piece of paper. Some very small black specks of undetermined origin were evident in the contact vicinity but not on the contacts proper, before cleaning. The contacts appear to be gold or gold flashed, judging from their color. The reason the contacts caused a problem only on diode No. 2 is probably due to the fact that the relay is normally spring closed in the diode No. 1 position and the contacts touch on edge in that position, resulting in much higher pressure per unit area of contact than in the diode No. 2 position where the contacts touch on a larger surface area. I believe that a mercury wetted reed relay would give superior reliability in this application. Also, attention should be directed towards the connectors for this signal path as it was discovered during the course of the trouble-shooting that a worn or dirty BNC connector could cause even worse problems than those caused by the relay.

Rewrite this memo to accomplish the following goals:
a. Use three-part organization.
b. Break the report into sections and use headings and lists where appropriate.
c. Cut excess wordiness.

5. Write a memo to the author of the memo in the previous exercise. Assume you are his/her supervisor and are tired of getting memos that are difficult to read and that waste your time. Explain what is wrong with the memo and make specific suggestions for avoiding the problem in the future. In your memo you should consider establishing a set of guidelines or an outline for future trouble-shooting and repair reports.

6. Write a memo to justify the raise described in the following:

Jane Bingham, your supervisor, has advised you that she has obtained for you a minimal 2% merit raise instead of the 5% you feel is justified. Although she feels she has done all she can, she will try again if you will write a memo stating your case. Because the committee meets today and won't meet again for three months, you have just one hour to prepare.

As you try to organize your thoughts, make a list of all the reasons for a better merit increase:

• always come to work on time.
• spearheaded department productivity study that resulted in 15 percent improvement.
• loyal to the company and have supported company social events.

- have voluntarily worked overtime on important proposals to meet submission deadlines.
- always attend group and department meetings.
- get along well with your supervisor and other employees.
- are careful about not taking excessive break time.
- filled in for your supervisor during her annual leave, and supervised production of two successful proposals that resulted in construction contracts worth $4.5 million.
- served as treasurer for the employee flower fund.
- established the new computer data file listing details of all proposals written by the company in the past twenty years; this has significantly improved preparation time of new documents.

Choose the best reasons from this list, and write a memo that will sell the committee on a 5% merit raise. Because the original listing is "off the top of your head," you will probably need to do some cutting, shifting, and revising before you write your memo.

7. Confusion exists in your department about what are reasonable and proper travel expenses to include in travel reports. Write a memo establishing guidelines for the following:

- mileage to and from the airport.
- total expense per day for three meals.
- number of authorized phone calls and how to report.
- laundry expenses.
- hotel/motel expenses.
- miscellaneous items.

Explain which items require receipts, how to document or justify necessary variations from the guidelines, and how soon the report must be filed after returning from a trip.

8. Your department has purchased some new equipment. Because no one else is familiar with its operation, you have been asked to write a memo explaining step-by-step how to load and operate the machine. Choose a machine with which you are familiar, such as a 35 mm camera, photocopy machine, computer printer, or turntable. Limit the subject to something simple such as loading film in the camera, putting paper in the copy machine or printer, or changing a needle in a phonograph. Write a maximum of one and one-half pages.

PART IV
WRITING RESEARCH PAPERS

12

Proposal for a Research Paper

189

In industry, a proposal is often a written response to a request from another company to have work performed. The requesting company (buyer) sends out *requests for proposals* (RFP's) to a list of companies (sellers) qualified to perform the type of work needed. Such work may be to build a dam, a road, a railroad, a ship, a fleet of airplanes, or a radar system for the U.S. Navy. It could also involve a service such as managing construction or improving the security at a small research facility. One business needs work performed and notifies other companies via an RFP about the need.

Companies receiving the RFP review it carefully to see if they want to offer the requested service. If the management of a company decides to bid for the job, the company must respond with a proposal or an explanation of how it will accomplish the task, who will do it, and at what cost. The result is often three separate proposals—The Technical Proposal, The Management Proposal, and The Cost Proposal.

Because this method is the only way some major aerospace, contracting, and manufacturing companies get business, writing effective proposals is important to them. For example, several years ago three aircraft companies spent millions of dollars and employed thousands of people to design an aircraft and write proposals to sell the idea to the Air Force. Only one company was successful.

The RFP takes many forms. For a major project it may entail hundreds of pages; for a small earth-filled dam on a ranch, it may be a one-page letter. Instead of calling the requesting document an RFP, some writers ask for a Request for a Quotation (RFQ), a Request for a Bid (RFB), or an Invitation for a Bid (IFB). Although they may vary substantially in format, the purpose is essentially the same: someone needs work performed according to certain conditions.

On major projects, the RFP details the exact form that the proposal must take, even providing a complete outline. This is particularly helpful to the customer (buyer) who wants to compare four or five proposals. For the writer(s), success often depends upon how carefully and fully the proposal meets all the specifications of the RFP.

This chapter does not tell you everything you need to know about writing proposals for big business; instead, it focuses on a practical writing project in which you write a proposal for a major research paper or report (you will write this longer paper when you study Chapter 13). The first section is a detailed RFP outlining what is wanted; this states the work to be done. The second section explains how to choose and restrict a subject and gather preliminary data. The concluding sections contain a model of student writing, review questions, and a writing assignment.

RFP—STUDENT RESEARCH PAPER

This section is a Request for Proposal (RFP) for the research paper you will write in this course. It is similar to those found in industry. And, as in industry, the success of your proposal will depend on how responsive it is to the RFP.

REQUEST FOR A PROPOSAL

Proposals for a major student research paper are currently being solicited for the English XXX Technical Report Writing class at XXXXX College. All proposals should meet the requirements outlined below. The deadline for submittal is X:XX A.M., (Month and day), at the English Department office.

General Description of the Research Paper

The research paper that the proposal will describe should contain 3,000–4,000 words in the body, supported by a title page, letter of transmittal, table of contents, list of tables and illustrations, informative abstract, appropriate documentation, and appropriate visuals. The document will be submitted in two copies, one bound. All work must be neat and professional in appearance.

The subject of the research paper is open as long as it involves interpretation of data. Every paper must be based upon a question that needs to be answered: for example, *Can a Craftsman Make a Living Producing Custom Furniture and Other Specialty Wood Items?* Conclusions, the answers to the question, will be inferred from data gathered from published sources and/or primary sources such as personal laboratory or field research.

The following information is a description of the proposal that must be submitted and accepted before authorization to perform the work (writing the research paper) can begin. The proposal is a statement of what you think will be your subject and approach; it is your preliminary estimate of what you will write.

Description of the Proposal

The proposal, organized in the standard three-part format, is written in paragraph form except for the outline and bibliography. It must convince your instructor that you have performed preliminary research, that you have a narrow but workable subject, that you have asked a question that can properly be handled in the space available, that you have a workable outline of what you will write, and that you have found sufficient data to support it adequately. The length of the proposal will be 500–750 words.

Here is a sentence outline that suggests the content and structure of the paper:

(Continued)

Outline

   I. Write an Introduction.
      A. What is the subject of this proposal?
      B. What is the purpose of this proposal?
      C. What is the scope of this proposal?
      D. What is the plan of organization of this proposal?

  II. Define the Problem.
      A. State briefly the general subject being investigated in the proposed research paper.
      B. Indicate your basis for restricting and focusing this subject.
      C. In just one sentence, state clearly the restricted question or problem you plan to research.
      D. Indicate the purpose of your paper—what it should do for you, to whom it is written (your intended audience), and what use you think the paper will have for your audience.

 III. Analyze the Problem and Your Approach.
      A. What are the subquestions that you plan to ask?
      B. What will be your standards of judgment in interpreting the data?
      C. How do you plan to organize the document? Write a tentative outline and include it in the proposal.

 IV. Explain How You Will Meet These Requirements.
      A. How will you use writing techniques taught in this text?
      B. How will your subject lend itself to graphic support?
      C. Is there sufficient material to perform a reasonable investigation? List at least ten sources that you plan to use, including specific titles and publication data. A reference to a periodical article should contain the author, title of the article, name of the periodical, month and day of publication, year, and page numbers. A book reference should cite the author, title of the book, place of publication, publisher, and year of publication.

Remember, the proposal is a description of what you plan to do in the research paper—it is not the report itself; therefore, it will contain little, if any, of the research data.

WRITING THE PROPOSAL

Choosing an appropriate subject and restricting it properly are the keys to success in writing the proposal and the research paper that follows. Try to visualize a funnel (see Figure 12.1)—wide at the top and narrow at the bottom. If you were to shine a flashlight through it, only a very small circle of light

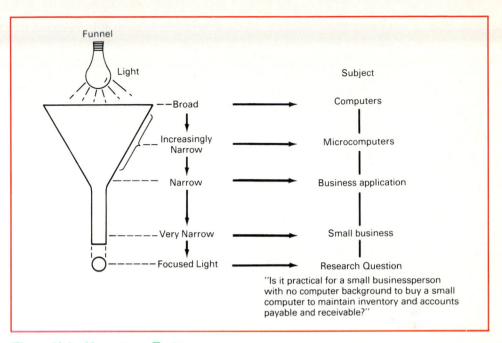

Figure 12.1   Narrowing a Topic

would come through the small end. Narrowing a topic to a specific subject is similar to shining a light through a funnel: you start broadly and gradually narrow the subject until you have something specific enough (like the small circle of light) that you can examine in great detail.

For example, suppose you are interested in computers and want to write a research paper about them. *Computers*, an extremely broad subject, is totally inappropriate for a research report until the topic is narrowed (see Figure 8.1). What *about* computers? What *kind* of computer? Ask yourself these and other questions as you search for a properly limited subject. You can narrow the field considerably by restricting *computer* to microcomputers with a memory of 256K or less. Now start asking questions—i.e., what will the computer be used for? what is an angle for me to research? You might try a discussion about computers in the home or in a small business. You should eventually arrive at a specific question such as this:

> Is it practical for a small businessperson with no computer background to buy a small computer to maintain inventory and accounts payable and receivable?

This question is narrow, like the small end of the funnel. If possible, you might follow up this question by actually visiting several local stores to see if one of the owners is asking the same question. If the answer is yes, you have an interested audience and a source of data. If not, you can still research the subject by gathering data from other sources. Remember, your topic *must* involve an interpretation of data, *must* reach some sort of conclusion, and *must* be framed in one good sentence.

For many subjects, there will not be a single source that answers your question. You will have to gather the data and make a recommendation. For example, one or two good books on solar energy can give you all the facts you need to understand theory and application in this field. However, if you direct your research question to a specific geographic area such as New England and ask, "Is it economically feasible to retrofit a frame building in the Boston area with solar panels to reduce heating bills?" You *personally* will have to check a variety of sources, and you *alone* will have to decide what is feasible. This is the essence of interpretation of data!

Choose a subject that interests you. It could relate to your major, a hobby interest, or a subject you would just like to know more about. Part of the excitement of research is what you learn: you'll become an expert on your topic!

The following research questions show the diversity of subjects possible for research topics.

1. What is the total cost of operating major appliances in the average home in (city, state)?
2. What are the effects of Saturday morning television advertising on children?
3. What is the economic feasibility of switching a modern all-electric home to a wood-heated home?
4. What is the effect of jogging on mental health?
5. Is a motorcycle a practical alternative to a car for commuting in (state or region)?
6. What is the feasibility of implementing the bovine embryo transfer process at the Lazy J ranch?
7. Is it economically feasible to build a small hydropower plant using impulse turbines to supply electricity for a farm?
8. Does mainstreaming the educable mentally retarded really work?
9. Is it more economical to make or to buy clothes?
10. Is it economically feasible to retrofit an existing home with solar panels for space heating in (city, state)?
11. What is the role of competition as a factor in Rocky Mountain bighorn sheep management?
12. Are homemade foods more economical and nutritious than convenience foods?

Be sure to formulate your topic into a question as a means to help you focus upon a limited, specific point. The question format also leads you to an interpretation of data, because you have to answer a specific question.

Narrowing the topic and doing preliminary research go hand in hand. Often you do not know enough about a topic to do an effective job of narrowing it. Do some preliminary reading in encyclopedias and periodicals to broaden your background enough to enable you to ask pertinent questions.

The RFP specifies a bibliography of at least ten entries. After you have narrowed your topic to a basic question, check the card catalogue, the *Reader's Guide to Periodical Literature*, specialized periodical indexes, specialized bibliographies in the reference collection, and outside sources to be sure sufficient material is available. Consult a reference librarian for help in focusing your subject and in finding additional sources. List at least ten in your proposal as proof that data is available. If you are unable to find sufficient support, it is wise to change subjects or at least your approach to the subject. Because you will have limited time to write the research paper, you simply must have sufficient readily available data from the beginning.

By its very nature, the proposed research paper may differ somewhat from the research paper you ultimately will write; this is normal. Remember, the proposal is your best estimate of what *you think you will include in your research paper*. Sometimes as you explore the subject, you may need to narrow or broaden your approach.

The following proposal is a response to the RFP on p. 191.

MODEL   **THE FEASIBILITY OF CONVERTING A DAIRY OPERATION TO A FULLY AUTOMATED SYSTEM**

This document is a proposal for a research paper required in English XXX, Technical Report Writing. It clearly defines a research question concerning a small dairy operation and shows its soundness as a research subject. The proposal is divided into three parts: (1) definition of the problem, (2) an analysis and approach to the problem, and (3) explanation of how this approach meets other RFP requirements.

Definition of Problem

Advancing technology has aided many industrial and agricultural fields, including dairy production; consequently, the older systems of milking are becoming outdated while the new automated systems are helping dairy farmers increase milk production. Because of the high cost of automated systems, many dairy farmers have been "scared off" and have not looked into the advantages or disadvantages and the net gain or loss that would result from the changeover.

The proposed research paper will, therefore, investigate the following question: Is it economically feasible for a dairy farmer to convert to a fully automated system? It is assumed that the dairy has an established milking system already in operation and a herd of 45–65 cows. The paper will allow the dairy farmer to evaluate the existing system and determine if an automated system would be profitable.

Analysis of the Problem

The analysis included in the research paper will answer the following questions:

1. Would conversion allow one person to run the entire operation where two are presently required?
2. How much time will the new system save in milking and clean up, or will it save any?
3. What type of automated system is best, of the systems presently available?
4. How much would the changeover cost?
5. How soon will the new system pay for itself through increased milk sales?
6. Could the new system be installed in the present milking barn, or would remodeling need to be done?
7. How long would it take to install the new system?

The data received from the research of the preceding questions and others will be compared to reach a conclusion as to the feasibility of the conversion.

The proposed research paper will be organized as follows:

I. Introduction of Subject Problem
   A. Subject
   B. Purpose
   C. Scope
   D. Plan of Organization

II. Automatic System Basic Components
   A. Feeder
   B. Lock-in Stantions
   C. Pulsater Remover
   D. Timed Washer
   E. Waste Removal

III. Feeder
   A. Cost of Each Type
   B. Effectiveness
   C. Recommendation

IV. Lock-in Stantions
   A. Cost of Each Type
   B. Effectiveness
   C. Recommendation

V. Pulsater Remover
   A. Cost of Each Type
   B. Effectiveness
   C. Recommendation

VI. Timed Washer
   A. Cost of Each Type
   B. Effectiveness
   C. Recommendation

    VII. Waste Removal
        A. Cost of Each Type
        B. Effectiveness
        C. Recommendation

   VIII. Feasibility of Conversion
        A. Labor Savings
        B. Time Saved
        C. Remodeling Costs
        D. Time Needed for Installation

    IX. Conclusion and Final Recommendation

How Approach Meets RFP Requirements

This approach to the research topic will use the techniques taught in English XXX: classification, definition, and interpretation of data. It will also include pictures or drawings of all key equipment and modifications as well as several graphs showing the costs, gains, and losses.

    A preliminary examination of sources to support the proposed research revealed many sources, including the following:

1. Bath, Donald L. "New Approaches to Dairy Waste Management." In *Dairy Science Handbook.* Phoenix: Agriservices Foundation, 1972.
2. Doan, M. K. "Air Flow Utilization in Milking Parlors." *Journal of Dairy Science,* 65 (1982): 835–42.
3. Kruse, Marvin L. "Dairy Cattle Buildings and Equipment." *Dairy Science Handbook.* Phoenix: Arizona State University, 1968.
4. *Milking Parlor Systems.* Morton, IL.: Clay Equipment Corporation, 1982.
5. Nonce, Paul, Representative of Snake River Dairy Systems. Personal Interview. Ririe, ID, 24 March 1985.
6. Rigby Chapter of Future Farmers of America. *Dairy Cattle Milk Production,* Rigby, ID. (No Date)
7. Sistler, Fred E. "Milking and Feeding in a Confinement-Stall Barn." *American Society of Agricultural Engineering Transactions,* 23 (Mar. 1980): 419–22.

This proposal is responsive to the RFP in nearly all ways. In the first paragraph, second sentence, the author avoids possible confusion of the proposal with the research paper. Of course, the subject, purpose, scope, and plan of organization constitute most of this paragraph. The second paragraph gives the reader perspective and background on the problem. The third paragraph clearly states the research question and then clarifies the author's assumptions—"established milking system" and "herd of 45–65 cows."

In the analysis of the problem, the author clearly poses some of the subquestions that must be answered in order to answer effectively the major question of feasibility. The outline has sufficient detail to show the order and form the research paper will take. The author ties the paper to the writing techniques taught in class and gives an overview of the planned visual support. The bibliography shows that sufficient data is available to perform the research.

REVIEW QUESTIONS

1. In industry, what is a *proposal?*
2. What is an RFP, RFQ, RFB, and IFB?
3. Why are carefully written proposals important to many industries?
4. What is the basis for evaluating the responsiveness of a proposal?
5. How many words or pages should your research paper be? (See RFP)
6. Although the subject is open (you may choose), it must be a(n) _____. (Fill in the blank.)
7. What are the keys to writing a successful proposal and research paper? (See Section 12.2)
8. Where can you find data to help you narrow your subject and formulate a worthy research question?

EXERCISES

1. The following subjects are too general for a student research paper. Narrow each topic until you have a subject limited enough for a 3000–4000-word research paper, and then write each as a specific research question.
   a. computers
   b. engineering
   c. solar energy
   d. fast-food franchises
   e. paints
   f. front-wheel drive automobiles
   g. foreign policy
   h. physical fitness
   i. jogging
   j. artificial hearts
   k. sugar substitutes
   l. allergy
   m. interest rates
   n. sewing machines
   o. motorcycles
2. *Writing Assignment.*   Unlike the others, this chapter includes the writing assignment near the beginning, in the request for a proposal (RFP). All

details of format, length, and choice of topic are discussed there and on p. 191. Write a proposal in response to the RFP. Your success depends upon how responsive your paper is to the requirements listed in the RFP.

Remember, this is a proposal for the major research paper that you will write in Chapter 13.

13

The Research Paper

In college, you are often assigned to write a research paper based upon published and/or primary sources. Research papers are also a feature of scholarly writing in technical publications. Pick up virtually any technical journal and you will find that nearly all articles cite original and/or scholarly research published in other journals; these articles often include documentation (footnotes or textual citation and bibliography).

Chapter 12 discussed the basic characteristics of the research paper assignment; in your proposal, you presented a limited subject based upon a question requiring interpretation of data. You also wrote a tentative outline and thought through the basic plan for your paper.

The present chapter picks up where Chapter 12 stopped: it presumes you have written a proposal and that now you will finish the project by writing the research paper. This chapter is divided into two major sections: *Preparing to Write* and *Writing the Research Paper*. The first explains how to gather data, how to write a working bibliography, and how to take notes. The second discusses understanding the scientific method, avoiding plagiarism, documenting the paper, and organizing the material. The chapter concludes with some details about handling writing problems, and includes review questions, and exercises.

PREPARING TO WRITE

Before you attempt to write the paper, systematically search all available sources to make a list of books, periodical articles, pamphlets, and other works that may be useful. Record each potential source of information on a 3″ × 5″ card; when you have collected cards on all the available sources you have a *working bibliography*.

Once you have gathered a working bibliography, systematically search the books and periodical articles for information that fits your tentative outline. On 3″ × 5″ or larger note cards, record data that appears to be useful. Once you have collected all the data available, you are ready to write your paper.

The previous two paragraphs provide a general overview of preparing to write. More specifically, this section explains what you need to know to (1) gather data, (2) write a working bibliography, and (3) take notes.

Gathering Data/Writing a Working Bibliography

Because you chose a narrow, restricted subject, you should be able to gather in-depth information on it. Your first stop will probably be the card catalogue.

The card catalogue, a listing of all works in the library classified under author, subject, and title, is usually found in a conspicuous location. In some libraries you must look for a microfiche reader or a computer terminal as shown

Figure 13.1 Computerized Card
Catalogue

in Figure 13.1. At the beginning of your research, use the catalogue as a subject index. If you are interested in solar energy, look for *Solar Energy*. Cross-references will lead you to other materials. For example, *See also* cards on solar energy may refer to *Alternative Energy Sources, Energy, Solar Power, Photo-Voltaic Cells*, and others. Check these entries and record the publication data about each book, including call number, on a 3″ × 5″ card—list only one book on each card.

The 3″ × 5″ card, usually called a *bibliography card*, should include the following data:

- author
- title
- edition (if included)
- translator (if included)
- editor (if included)
- place of publication
- publisher
- year of publication
  (Include the call number to save time if you must refer to the book later to verify information.)

Figure 13.2 is an example.

```
William, Peter J.

    PIPELINES AND PERMAFROST
    PHYSICAL GEOGRAPHY AND DEVELOPMENT
    IN CIRCUMPOLAR NORTH

    Longman Group Limited, London, 1979

    Call # TN 880.5 W55
```

Figure 13.2   Bibliography Card: Book

Once you have written bibliography cards, one for each book listed in the card catalogue, check the periodical indexes; they are particularly helpful in providing up-to-the-minute information. A good place to start is the *Reader's Guide to Periodical Literature,* which indexes general newsstand periodicals. Smaller libraries tend to have more of these periodicals on hand than the specialized journals. Again, materials are classified under author, subject, and title. Try to concentrate on the last five to ten years for each periodical—these references are most up-to-date and most likely to be available. Again, write down all information about a particular article on a $3'' \times 5''$ card:

- author
- article title
- name of periodical
- volume
- number
- month
- day
- year
- page numbers

Figure 13.3 is an example of a note card for a periodical.

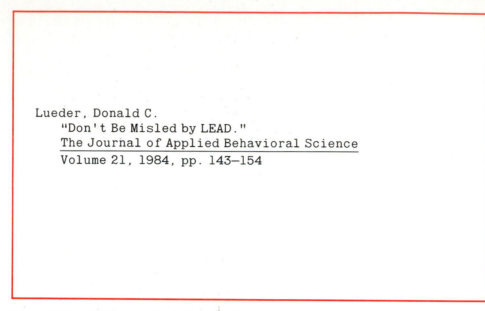

```
Lueder, Donald C.
   "Don't Be Misled by LEAD."
   The Journal of Applied Behavioral Science
   Volume 21, 1984, pp. 143—154
```

Figure 13.3   Bibliography Card: Periodical

Remember that at this point in your research your purpose is to gather a complete listing of all sources that may have information related to your subject; you will investigate them thoroughly later.

Look around the same area in the library where you found the *Reader's Guide,* and you will probably find other periodical indexes such as *Applied Science and Technology, Engineering Index* or the *Business Periodicals Incex.* Many libraries have specialized indexes relating to professions; among these is the *Nursing Index.* Write a new bibliography card for each fresh source found in these indexes.

The reference collection is also an excellent source for basic data. You probably looked here when you were doing your preliminary survey of data for your proposal. Specialized bibliographies—entire books that have nothing in them but lists of other books available on specific subjects—are also available here. At the end of many articles in the encyclopedias are short bibliographies. Sometimes these make excellent sources, so once again make a bibliography card for each entry that you feel may relate to your subject and outline.

Talk to the librarian about other sources, such as a computerized data bank or clipping service. Often the librarian can save you considerable time by pointing you in the right direction.

If your research topic and question are properly limited, it should be possible to include virtually every source that may relate.

A word of caution is now in order. Scientific and technical writing requires careful, precise use of information. When you write your paper you will have to use footnotes or textual citations and bibliographical entries for all the information and ideas you obtain from sources. You must begin now to be careful and accurate in listing the information on your bibliography cards.

Probably the greatest advantage of gathering a working bibliography by listing one work per bibliography card is the ease with which you can alphabetize the cards later when writing the bibliography (sometimes called *Works Cited*) for your paper. At that time you will simply shift the cards around, throwing out those that were not helpful, until you have the right order. When you include the bibliography in the research paper it is usually in alphabetical order.

Taking Notes

Once you have your working bibliography completed, systematically research each of the books or periodical articles listed for data relating to your tentative outline. Record this information of 3″ × 5″ or larger note cards also.

To take good notes, record just one idea per card. Later, when you are ready to write, you can arrange your cards in an effective order to support your outline. If, however, you have two or more ideas on one card, then that card will need to be in two or three places at the same time: chaos results.

Because you will have to document all the ideas and information used in your paper, you have to know the source, including page number, of every bit of data. To record this accurately, you may use several different systems of notation. One of the most common is to list the author's last name and the page number on the bottom of each note card. All of the other publication data needed, of course, you have recorded on the bibliography card! If there are several books by the same author, or several by authors with the same last name, add the first few words of the title of each work (in addition to the author's last name) to distinguish it from the others.

Another good way to accomplish the coordination of note cards with the source is to number or letter each bibliography card and then place the number of letter with the appropriate page number on the note card. See Figures 13.4 and 13.5 for examples of the two forms.

Because full data on *Ski Cross Country* by M. Michael Brady and Lorns O. Skjemstad is available on the bibliography card, the card only needs the notation *Brady, p. 53*. If other books in the bibliography have authors with similar names, add just enough data to make it clear which one is being quoted or referred to on the note card.

In the number/letter system of card correlation, you must write the number or letter in a conspicuous spot on your bibliography cards, clearly identifying it. On the note card, then, all you have to do is record the number/letter from the bibliography card and follow it by the specific page number indicating where the borrowed material is found—*1, p. 53*, or *A, p. 53*. Obviously, if your bibliography includes more than twenty-six references, you should number the cards instead of using letters.

EQUIPMENT AND CLOTHING—2.0

"There is no single 'right' type of equipment or one 'right'
type of clothing for ski touring and cross-country ski rac-
ing."

Brady, p. 53.

Figure 13.4   Sample Note Card: Author/Page System

EQUIPMENT AND CLOTHING—2.0

"There is no single 'right' type of equipment or one 'right'
type of clothing for ski touring and cross-country ski rac-
ing."

A, p. 53.

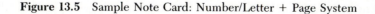

Figure 13.5   Sample Note Card: Number/Letter + Page System

Perhaps you are curious about the *2.0* notation listed after the title of the cards in Figures 13.4 and 13.5. This number refers to part two of the author's tentative outline—such as the one you wrote for your proposal. Later, when you start to write your paper, such a code system will tie your notes directly to one part of your outline. Once your research is complete, you can easily spread all cards with 2.0 or 3.0, and so on, on a table and then shift them around until you have the most effective sequence. Then you write! It is also good practice to give each note card a title so you can recognize its content quickly.

At this point you are probably wondering what to put on the note cards. Most of the time you will summarize in your own words what you read. A *summary* is usually much shorter than the original material. Sometimes you may want to *paraphrase*, to restate in your own words without attempting to reduce the length. The summary and the paraphrase are helpful to you as the reader/writer because they help you understand the material. Occasionally you may want to write a *précis*, a concise summary that maintains the emphasis, tone, and proportion of the original. Although it has a place in report writing, most students rarely use précis. Indicate *direct quotations*, material copied word for word, on your note cards with quotation marks. Use direct quotations for no more than 10 percent of your research paper. A higher percentage indicates that you have not assimilated the material and controlled it.

In summary, then, before you are ready to write your paper, you must make a working bibliography of all the sources you can find, and then systematically examine each for material that relates to your outline. When you find pertinent material, place it on a note card and indicate its exact location by making careful reference to the page number on the bottom of the note card. Every note card should have a source and page reference. As you make notes you will write them in summary, paraphrase, précis, or direct quotation.

WRITING THE RESEARCH PAPER

This section discusses the scientific attitude, how to avoid plagiarism, how to document a paper, and how to write the paper, from title page to appendix.

Scientific Attitude

Your research paper should carefully weigh the evidence pro and con and then arrive at a carefully reasoned, conservative position. The tradition of scientific writing often requires statements such as "The data appears to indicate . . ." instead of the more definite "This paper shows. . . ." As a student writer, try to be neutral in researching your subject. Whether your research supports or opposes a particular subject or simply comes out some place in between, the subject is still news. You perform your research on your carefully limited subject, and then accept the results whatever they may be.

The scientific attitude, as mentioned previously, also requires you to document carefully information borrowed from a source. The idea is to make your paper complete enough so that a reader interested in reading your sources can easily find exactly the same page you studied as you wrote your paper. You make this possible by recording data accurately for the bibliography and then taking accurate notes.

Plagiarism

Plagiarism is defined as using the words or ideas of someone else as your own. If, for example, you borrow material from a source—copy it word for word—and do not tell the reader, you are plagiarizing. If you borrow an idea that is not common knowledge and use it in your paper without documenting it, this is also plagiarism, or, put more bluntly, *stealing*. This means that when you paraphrase or summarize someone else's words or ideas, you have an obligation to credit the source of the material. Plagiarism refers to *ideas* as well as the *exact words* of another!

As the author of a research paper, you know whether the material you are using came from you or from a source. A good rule of thumb might be this:

If information comes from a note card, document it.

Of course, if the material used in the paper is common knowledge—that is, something your typical reader might already know—documentation is not required. For example, most people know that Richard Nixon served as President of the United States. However, when you start to talk about a particular law he sponsored on a particular date you are moving beyond common knowledge.

In a practical sense, then, much of your research paper will require documentation. If you have not yet mastered the advanced concepts involved in your major field, you have only general knowledge about many specialized concepts. Once you have done all you can to integrate data, to put in your own interpretation and commentary, your only obligation is to be honest; do not worry about having too many notes!

Documentation

Documenting means to indicate the source of the words/ideas that you have borrowed. For many subjects this has traditionally meant writing footnotes and a bibliography, usually according to a pattern prescribed by a professional organization such as the Modern Language Association (MLA). Because these organizations publish journals with particular subject interests, several have established documentation formats that they feel particularly fit their subject needs and have required authors publishing in their journals to follow them. Another well known system is the APA style based upon the Publication Manual of the American Psychological Association. Various systems emphasize different information in their textual citations: some emphasize author, title, page; others stress author and data; others may use only a single number. This text favors the MLA format.

The MLA format has recently (January 1984) been modified to eliminate footnotes at the bottom of the page or in a listing at the end of a work. Instead, this format favors textual citations. Under this system, if you mention an author in the text of your paper, all you need in addition is the specific page number of the source placed in parentheses before the period ending the cited section. In this system, if you cannot conveniently mention the author's name, you insert it in the parentheses and follow it by the period ending the sentence. If there are several authors with the same last name or you have several articles by the same author, you can distinguish among them by adding a key word from the title. To complete the documentation, place a *Works Cited* page, a bibliographical listing in alphabetical order of all the sources you used, at the end of the research paper. The following illustrations show the format for textual citations based on three examples from a student paper.

Guidelines for Textual Citation

Example 1:   Glen L. Witt says that Port Orford Cedar is the preferred species of boatbuilding cedars (37).

Mentioning the author's name at the beginning of the citation and then placing the page number at the end clearly marks the borrowed idea.

Example 2:   Lofting, expanding or redrawing the lines of a boat to full size, can be a complex process. Witt says, "The main reason for the lofting process is so the builder can pick up the hull contours, transfer these to the members that will be used to build the boat. . . . (54)."

In Example 2, the student author cited a direct quotation of four lines or less. Notice how the author integrated the quotation into the text. The dots at the end of the sentence indicate the student author did not use all of the sentence.

Another author clarifies how a marine transmission differs from the transmission in a standard shift automobile:

Example 3:   In boating, the boat must be able to have quick and positive control

between forward, neutral, and reverse directions. If an automotive

type clutch were used, there would be a considerable time lag

between changes in direction which could be hazardous, especially

> when operating in crowded conditions. Furthermore, the ability
>
> to shift from forward to reverse, and vice versa, is in effect, the
>
> breakes of the boat. Consequently, most boats use and require
>
> some sort of marine transmission.
>
>   (Witt 40)

A quotation over four lines long is indented ten spaces on the left and double spaced. No quotation marks are needed. The textual citation is placed close to the right margin below the text as indicated in Example 3.

The new MLA format and others such as the APA make good sense to many authors. Older systems seem unnecessarily redundant because the data in the footnote and in the bibliography are virtually identical. Printers favor textual citation because of the lower cost and greater ease in printing. Students who used to run out of space on the bottom of carefully typed pages before they could squeeze in all the required footnotes also favor textual citation!

One major drawback of textual citation is the limited information readers have on the page they are reading: if they want to find out more about the reference, they must turn to the *Works Cited* section at the end of the paper. Many readers, apparently, are willing to put up with this inconvenience.

The following section explains how to write works-cited entries for commonly used documents. For additional examples, study the bibliography of the sample research paper or an MLA Style Guide.

Guidelines for "Works Cited"

1. Book—Single Author

   Orr, W. I. *Radio Handbook*. 20th ed. Indianapolis, IN: Editors and Engineers,
       1985.

   Typical order is author, title of book, edition (if other than first edition), place of publication (include state if the city may not be well known), publisher (Editors and Engineers in this example), and the copyright date. The title is underlined in typed text. In MLA format the entry is double spaced.

2. Book—Multiple Authors

   Baumeister, T., E. A. Avallone, T. Baumeister III. *Marks' Standard Handbook
       for Mechanical Engineers*. 8th ed. New York: McGraw-Hill, 1985.

   If only two authors are mentioned, use *and* and list the second in the same order as in the example. The full name of the publisher is McGraw-Hill Book Company; however, the shorter form is preferred for documentation.

3. Encyclopedia

Lawrence, T. G. "Animals." *The New Book of Knowledge*. 1980 ed.

4. Periodical—Monthly

Scott, David. "Optical Data Cassette Stores 2.5 Million Pages." *Popular Science*, December 1985, 21.

This is the format for a common newsstand magazine such as is listed in the *Reader's Guide to Periodical Literature*. Typical order for a monthly periodical entry is author (if one is listed), title of article, name of periodical, month, year, and page. The title of the periodical should be underlined in typed text.

5. Periodical—Weekly

Skrzychi, Cindy. "Mass Transit Rolls Along—But Not Merrily." *U.S. News and World Report*, 9 Sept. 1985, 53.

The only difference between weekly and monthly entries is the addition of the day before the month.

6. Journal—Separate Pagination

Sloan, John W. "The Mexican Variant of Corporatism." *Inter-American Economic Affairs* 38.4 (1985): 3–18.

Journals are specialized periodicals typically published by a scholarly or professional organization. Most are listed in specialized periodical indexes. Some begin numbering pages with page 1 each month.

After the title of the journal—*Inter-American Economic Affairs*—is the *volume* (38) written as an Arabic number followed by a period. The *number* (4) is also written as an Arabic number.

7. Journal—Continuous Pagination

Lueder, D. C. "Don't Be Misled by LEAD." *The Journal of Applied Behavior Science* 21 (1984):143–154.

Most journals number pages consecutively throughout a given year or volume. This means that the first page in the June issue, for example, could be 509.

The only difference between this entry and the previous one is the *number*. With continuous pagination, this information is not needed for clear identification.

8. Interview

Graham, Richard, Physics Professor, Ricks College. Personal Interview. Rexburg, ID, 16 Nov. 1985.

Complete details of the MLA Style are available in various guides for writing research papers. It is a wise practice to find a good guide and use it for all your writing. It is also important to remember the philosophy behind documentation: to give the reader a clear idea of exactly where you found your information. Fall back on this philosophy when your guide, as will certainly happen, fails to include an example of the particular document you are trying to cite.

Organization

The research paper as discussed here is an extended interpretation of data paper such as is described in Chapter 9. Three-part organization is standard, as it has been in all writing assignments in this text. The discussion of outlining on pp. 23–29 is particularly important as review material. It explains how the outline gives signals to write introductions, headings, transitions, and content. To succeed, you must carefully organize your paper with clear signals to readers—do not let them get lost. This means the general introduction to the body of the paper should give an overview of the entire paper, including a statement of the subject, purpose, scope, and plan of organization. Each section and subordinate section of the paper will do much the same thing but in briefer form. This textbook uses the same form.

The following discussion lists all the specific sections you should include in your paper, discusses each briefly, and provides a model from a student research paper. Spacing requirements, use of headings, page numbering, and other features are indicated in the discussion and in the model. Consider this text as the style guide for writing your paper.

Your paper should include the following:

- title page
- letter of transmittal
- table of contents
- list of tables and illustrations (optional)
- abstract
- body of paper (ten-plus pages)
- appendix (optional)
- endnotes (placement and need depend upon format chosen)
- works cited

Title Page

The title page is the reader's first introduction to the paper and its subject. Make a good first impression by making it neat and functional. Use the research question as the basis for writing the title. Titles to technical papers vary widely, but are often several lines long. This gives the author an opportunity to identify the subject clearly and to use key words that may be useful in indexing the document.

The title page should include at least the following information: the title,

Report on

THE FEASIBILITY OF WIND–GENERATED POWER ON THE
SOUTHERN OREGON COAST USING A NIMBUS
WIND ENERGY CONVERSION SYSTEM

by

Leonard Clayton Dahle

Submitted to
Dr. William D. Conway
English 216
Ricks College
Rexburg, ID
April 1, 1983

Figure 13.6   Model: Title Page

author, date completed, to whom the paper is submitted, and the school where
submitted. The centered format used is both neat and attractive. Use Figure
13.6 as a model for your title page.

Letter of Transmittal

The letter of transmittal should use one of the standard formats mentioned in Chapter 10 (block, semi-block, or full block). It is a letter from the writer to the person who will receive the report; college students, for example, would address it to the instructor who assigned the paper. In the business world, a preface is often substituted for the letter of transmittal. This is because the report might be read by a wide audience—a preface has broad appeal.

The content of the letter is pretty well standardized:

1. The first paragraph usually mentions the occasion for the report (in business this may be a contract) and cites the title. The first paragraph in Figure 13.7 is a typical example.
2. The second paragraph may explain the purpose and scope of the report and provide general background. Content varies widely: you can include key data if you feel it should be emphasized. Sometimes an additional line or two acknowledge the name(s) of anyone who provided significant help in preparing the paper.
3. The letter closes cordially with a statement such as this: "I hope this report will prove satisfactory."
4. After the "Sincerely" or "Sincerely yours," be sure to sign the letter.

For a complete discussion of business letters, read Chapter 10. Pages 143–166 will be particularly useful in helping you construct your letter of transmittal. Figure 13.7 is a model of the letter of transmittal.

Table of Contents

The table of contents is a polished version of your outline. It should accurately represent the exact form of every heading in the paper to at least the third level. This means entries in the table of contents must be identical to headings in the paper, except in underlining and numbering to the second and third level levels. Although the number (1.1.2) or letters/numbers (A, 1) are part of the headings in the paper, they are deleted in the table of contents to improve readability. Not every textbook follows this practice; it is a matter of format style. For your paper, this textbook establishes the style.

The table of contents should also include a *List of Figures* as the first entry if you use more than four figures or illustrations in the body of the paper. For example, in Figure 13.9, *List of Figures* and/or *List of Tables* would be the first entries after *Table of Contents*. If you use fewer than four, this entry is

261 South 100 East
Rexburg, ID 83440
April 4, 1983

Dr. William D. Conway
English Department
Ricks College
Rexburg, ID 83440

Dear Dr. Conway:

In accordance with your instructions I have prepared the ac—
companying report entitled The Feasibility of Wind—Generated
Power on the Southern Oregon Coast Using a Nimbus Wind Energy
Conversion System.

Since the Arab Oil Embargo of 1973, the government has adopted
a policy encouraging the development of alternate energy resources. One of these resources, wind energy, has expanded
rapidly into a thriving industry with multi—million dollar
sales. Because of this success, one could assume wind energy
really has become a cost—effective energy source. I performed
the following research study to form a judgment on something
more concrete than assumption.

This report studies the feasibility of wind—generated power
under the most favorable conditions possible. It considers
the factors relative to an actual wind power production project and explores the physical, technical, and financial elements of wind energy conversion.

I hope this report will prove satisfactory.

Sincerely yours,

Leonard Clayton Dahle

Leonard Clayton Dahle

Figure 13.7 Model: Letter of Transmittal

not required. Again this is a matter of style. Basically, the table of contents
lists everything that follows it.

As you study the sample paper, you will notice lower case Roman numerals
on the bottom of the page. Research paper tradition requires their use on the

pages leading up to the body of the paper. The title page is page *i*, but this is not written on it; nor is *ii* written on the letter of transmittal. The table of contents, page *iii*, is the first to have the page number written on it. This same numbering system continues through all prefatory pages, including the abstract. In the body, where Arabic numbers are used, the first page numbered is page two. Traditionally, the first page is not numbered, even though it is counted. Study Figure 13.8 for a clarification of the relationships.

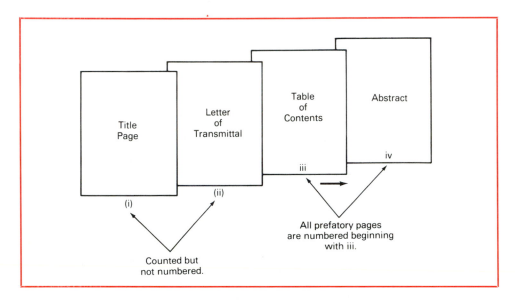

Figure 13.8   Overview: Prefatory Pages

```
                    TABLE OF CONTENTS

ABSTRACT        .   .   .   .   .   .   .   .   .   .   .   vi


1.0 INTRODUCTION      .   .   .   .   .   .   .   .   .   .   1


2.0 PHYSICAL CONSIDERATIONS    .   .   .   .   .   .   .   3
       Site Location    .   .   .   .   .   .   .   .   .   3
          Wind Speed  .   .   .   .   .   .   .   .   .   3
          Site Accessibility   .   .   .   .   .   .   .   4
          Grid Availability    .   .   .   .   .   .   .   4
```

Figure 13.9   Model: Table of Contents

(Continued)

Abstract

The abstract, a brief description or summary placed before the body of a report, gives readers an overview of the paper that follows. In a sense, the abstract is redundant because it may repeat information mentioned in the letter of transmittal, the introduction, and the conclusion of the paper. This is normal practice. The abstract is one of the principal means for readers to gain a rapid overview of a paper.

There are two types of abstracts. A *descriptive* abstract tells what the paper is about without revealing any of the findings—it sounds much like a statement of organization. The example in Figure 13.10 gives the reader an overview of the content, and little more.

The second type, an *informative* abstract, often has most of the elements of the descriptive abstract but goes well beyond this information by giving specific data and results. The informative abstract is preferred by many readers because it allows them to grasp the essentials of content without reading it. Contrast the example in Figure 13.10 with the one in 13.11.

For further clarification of the difference between the two basic types of abstract, study Figure 13.12.

```
This progress report summarizes the XXXXXXX Laboratory work
performed for the Reactor Development Program during January
1985 in the following research and development areas: Experi-
mental Breeder Reactor No. XI (EBR-XI), LMFBR Design Support,
Instrumentation and Control, Reactor Physics, Reactor
Safety, Energy and Environmental Systems, and Other Fast
Breeder reactors.
```

Figure 13.10   Example: Descriptive Abstract

The AIROS—IIA dynamic—simulation digital code was used to study the behavior of future ERB—XI cores containing predominantly oxide fuel. This study compiles a survey of the response of EBR—XI to hypothetical malfunctions of components. Previous studies, as well as operating experience at EBR—XI, have been with loadings of predominantly metallic driver fuel. The response to this fuel to a variety of operating conditions has been well documented. Future loadings, however, are expected to contain an increasingly higher proportion of experimental oxide fuels. The present study considers the response of predominantly oxide cores with and without doppler feedbacks. The study concludes that the future of dynamic behavior of EBR—XI with loadings of predominantly oxide fuel is predictable, safe, and well within the range of present operating experience.

—Courtesy Dept. of Commerce

Figure 13.11   Example: Informative Abstract

Here are a few suggestions for writing an abstract. Unless your writing requirements specify otherwise, use the informative style.

1. Begin with the research subject or question.
2. Use the table of contents as a source of data to write a brief summary of each section of your report.
3. Write in complete sentences to produce one paragraph that does not exceed one double-spaced typewritten page.

Figure 13.12   Relationship: Descriptive and Informative Abstracts

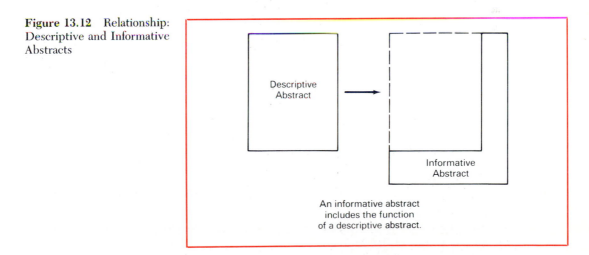

4. Make the abstract entirely independent of the body of the paper. In other words, do not expect the reader to know what is in the body of the paper and do not refer to illustrations and data within the paper. It is fine to mention that they exist, but do not expect the reader to turn to them to understand the abstract.

A good informative abstract covers the important points, conclusions, and recommendations of the report, including some of the significant details. It also suggests the proportion and emphasis of the original.

The following example, Figure 13.13, from the model research paper meets the specifications for a good informative abstract. It was a separate page in the research paper.

ABSTRACT

To determine the feasibility of wind-generated power on the Southern Oregon Coast, a location near Coos Bay, Oregon, was selected. Good accessibility and an available grid hook-up source made Cape Arago the site best suited for the study. Three important environmental factors were assessed and the following conclusions reached: 1) The moist coastal climate will be safeguarded against by painting and other maintenance. 2) Coastal storms will be prepared for by structure support and a system turnout mechanism. 3) The system's environmental impact will be approached conservatively. Site developmental considerations were explored and the total cost of site preparation is $24,280. The tower and system after evaluation were included in a combined cost of $120,000. The grid interface requires a fail-safe and frequency governor in the system at a cost of $4,000. Noise, aesthetic impact, and security measures were also considered, the latter costing nearly $1,000 to implement. Financial considerations include tax credits, financial resources, and cost effectiveness calculation. The major tax credit available is a federal credit of 40 percent up to $4,000 total. Three sources are possible for financing: government grants through DOE, the Solar Loan Program, and private business investments. A payback time of eight years is found by dividing total project price by the dollar amount of electricity produced in a year. It is concluded that wind power production on the Southern Oregon Coast is feasible when using a Nimbus wind energy system.

Figure 13.13   Model: Abstract

You may find it interesting to compare Figure 13.13 with the table of contents in Figure 13.9. The coverage of main ideas in the abstract seems very complete.

Introduction

The introduction to a research paper follows the same format as do all of the papers assigned in this text. Chapter 3 covers this in detail. As always, you need to get readers involved, to answer some of their questions. They must know what is going on. This is done with the standard introductory elements: statements of subject, purpose, scope, and organization. Sometimes you may need to add a little theory or history, depending upon your analysis of your audience.

Here are specific steps to copy in laying out the introduction to your research paper. (Figure 13.14 uses the following format.) Begin the title one-fourth to one-third page from the top of the sheet. Center the words *Report on*. Two space below it, center the title of the paper, underlined in capital letters just as it appears on the title page. Two spaces below it, center and underline the title—*I. INTRODUCTION*. Skip two more lines and begin the text. Maintain a pleasant balance of white space and text by using at least an inch of margin on all sides, with an inch-and-one-half to two inches on the left side to allow space for binding.

The model introduction in Figure 13.14 is clear and to the point. It carefully clarifies the subject, purpose, scope, and statement of organization.

Body

The body of the paper is the logical development of the subject specified in the introduction and outlined in the table of contents. Each new chapter of the paper starts at the top of a new page. And, as is outlined in Chapter 3, each section of the paper should begin with an introduction that tells the reader what the section contains. Although not as extensive as the general introduction that introduces the entire paper, the chapter introduction aims to accomplish the same purpose but for a smaller body of data.

In other words, your chapter introduction attempts to introduce the subject, purpose, scope, and plan of organization for just a single chapter. If one of the topics in the chapter is further subdivided into subtopics, you should also introduce the subtopics. The process of introduction applies to all levels of the paper. Review Figure 2.4 and the accompanying discussion if you need further information.

Here are guidelines for handling the headings in your paper. The discussion cites examples from the model research paper (see Figure 13.15).

1. The main heading (2.0 *PHYSICAL CONSIDERATIONS*) is underlined and centered on the page. Other than the underlining, it should be identical to the heading for Chapter 2 in the Table of Contents.

Report on

THE FEASIBILITY OF WIND-GENERATED POWER ON THE
SOUTHERN OREGON COAST USING A
NIMBUS WIND ENERGY CONVERSION SYSTEM

1.0 INTRODUCTION

During the last ten years, the United States has encouraged
development of domestic energy resources. One rapidly expanding resource with a bright future is the wind industry.
For many years when oil was inexpensive, wind power was discouraged because it was not cost effective. Now, with petroleum prices five times higher, advocates claim wind power has
finally come of age.

   The purpose of this paper is to analyze the validity of
recent claims that wind power is a viable cost-effective means
of power production. This study researches and develops an
actual wind energy conversion project, projects major costs,
and examines related considerations.

   Because the Southern Oregon Coast has favorable characteristics for wind power conversion, the scope of site locale
was limited to this area. The Numbus, due to its recent development and specifications, is used for wind power production
data. All other factors are general elements of wind power
conversion.

   Analysis is divided into three sections: the physical,
technical, and financial considerations of wind power production. These sections explore and provide information on each
level of project development, from initial site location to
tax credits and financing. Following the research data are
conclusions and recommendations.

Figure 13.14   Model Research Paper: Introduction

2. The second level heading (*Site Location*) is located on a line by itself, is
underlined, and has only the first letter of each word capitalized. Compare
it with the corresponding entry in the table of contents—here its subordinate
position is indicated by a three-space indentation. The *2.1*, indicating its
position in the outline, has been deleted to improve the readability of the
table of contents.

3. The third level heading (*Windspeed.*), located on the same line as some of the text, is underlined, followed by a period, and capitalized on the first letters only. Because three levels are about all a table of contents can handle easily, this is as far as the heading system goes. If you have fourth-level headings in your outline, experiment with a ten-space indentation or some other method to distinguish the heading from the other three levels.

Another format consideration concerns where to place the page numbers. They may be centered on the bottom or the top, or may be placed in either the upper or lower right-hand corner. Use whichever position appeals to you.

The sample pages that follow also contained visuals. Remember to follow the format established in Chapter 4 by placing visuals in boxes, by labeling them as *Figure 1*, and so on, and by giving them a title. Be sure to refer to each figure in the text of your paper. If a visual occupies an entire page, include it as an unnumbered facing page or place it in the traditional right-hand page sequence. If documentation is required, include it in the lower right-hand corner of the illustration, inside the box, usually by placing the author and page number in parentheses.

Study the way these matters are handled in the following pages (Figure 13.15). Notice that the final page contains a summary of the data in this chapter. Intervening pages have been deleted from this printing to conserve space.

2.0 PHYSICAL CONSIDERATIONS

Just as petroleum companies are limited to oil fields for crude oil production, wind energy conversion systems are restricted to those areas with the best physical conditions for maximum power generation. Factors to consider when evaluating potential areas include site location, environmental aspects, and site preparation. Following discussion of these considerations, a summary evaluates the findings.

Site Location

Finding the best possible site requires much time and effort, especially when determining the best of several good possibilities. Windspeed, the key aspect of feasibility, is weighed in light of other data such as accessibility, grid availability, and landscape. These criteria were used to select a test site for this study.

(Continued)

Figure 13.15 Model Research Paper: Body

Windspeed. Recent interest in wind-generated power has prompted several comprehensive studies that included the Southern Oregon Coast. One study by Oregon State University under the direction of Robert W. Baker determined which locations in the Pacific Northwest were best suited for wind energy conversion. The area with the greatest potential was Cape Blanco, a rocky headland 60 miles north of the California border. Here an average annual windspeed of 18.1 miles per hour was recorded with the wind coming out of the Northwest (64–65). Several other coastal stations reported similar favorable data. The conclusion drawn from the survey found the potential wind energy conversion region starting about 50 miles north of Cape Blanco, extending down along the Northern California coast. Based on these findings, two possible sites were chosen, Cape Blanco and Cape Arago (see Figure 1).

Headlands are well suited for wind energy conversion (WEC) system locations. Because they protrude into the ocean, many sources of wind disturbance such as forests, large rock formations, and mountains are absent. This allows the WEC system full influence of input wind force, and in turn, optimal power conversion.

Site Accessibility. Although headlands rate high in wind force utilization, they seldom are accessible. Their rocky make-up provides natural impediments for road construction and site development. At some locations, the U.S. Forest Service has helped solve this problem by building excellent service roads to aid in research and timber inventory. Because of enormous road construction costs, finding an already existing headland roadway makes a big difference in calculating cost effectiveness.

Grid Availability. Another factor in determining a site is proximity to utility lines. Hooking up to the power line network or "grid" is what makes revenue by wind energy conversion possible. As Gary Gardner of Utah estimated . . .

(Skips to page 12—"Fencing" in the sample table of contents.)

. . . over the entire site would take over 40 years and would cost $280 (Dahle). For the benefits gained, this additional cost is worth the investment many times over.

Fencing. For the security of the system equipment and for the general public's safety, a chainlink fence will be erected around the site. At the top of this eight foot fence, several strands of barbed wire will be strung to further deter would–

(Continued)

be vandals. With information supplied by Anderson Lumber Company (telephone interview), one can calculate the cost of the security fencing at $6.45 per linear foot. Multiplying this cost by the 720-foot perimeter comes to a total of $4,644. Then, after an additional $350 for the gate and corner braces, the total comes to approximately $5,000.

Summary

From wind power feasibility studies, two possible site locations are at Cape Blanco and Cape Arago on the Southern Oregon coast. Because Cape Arago is more accessible and has available power lines, it was chosen for the study.

Due to the coastal climate's adverse affect on machinery, painting and regular inspection will be used as preventative measures against corrosion. Coastal storms will be dealt with by assuring tower foundation stability and by outfitting the system with mechanical cut-out devices. It was determined that despite wind power's environmental advantage over conventional energy sources, continued consideration would be given to conserving the existing natural habitat.

3.0. TECHNICAL CONSIDERATIONS

With a site chosen and prepared to receive the tower and WEC system, this study centers on the technical elements of actual wind energy conversion. The following section analyzes tower and system specifications, grid interface, and other inputs related to wind power production. Following these, a summary of the data presented concludes the section.

Tower Specifications

Towers come in a virtually unlimited array of size and configurations. The purpose of a tower has the greatest influence in determining its height, constructional material, and unit cost. The tower chosen for this study is analyzed by these three criteria.

Height. Windmills International Diversified, the makers of the Nimbus, recommend a tower specially engineered for their system by James G. Smith (Lord 16). Careful research found this 75-foot derrick type tower to be the most stable while providing maximum elevation.

(Continued)

> <u>Construction and Material</u>. As stated above, the Numbus tower is similar to an oil derrick constructed of riveted steel beams 28 feet at the base narrowing to 6 feet at the top. Extra heavy rivets and beams are used to ensure structural strength for anticipated stresses caused by high winds. Cross braces stretch across horizontally and diagonally to further strengthen . . .

Conclusion and Recommendations

The conclusion may include a summary of the main points presented. Do not introduce new data at this point. Give readers clear signals that the end of the paper is near. Answer the question posed in the introduction and explain how you arrived at your answer.

The conclusion may include recommendations, or these suggestions could be under a separate heading, as they are in the sample paper. Study Figure 13.16, which contains the conclusion and recommendation from the sample paper.

Note: Sections 4.0 and 5.0 would each begin at the top of separate pages of double-spaced type in a student paper.

Appendix

Indicate the appendix on a single page, with the word *Appendix* underlined and centered. Although this page has no page number written on it, it is counted.

The appendix contains any information you feel is important support for the main paper but that is inappropriate for inclusion in the body of the paper. If you have collected considerable raw data, for example, a general summary and interpretation of it would be appropriate in the body. However, you would not wish to include all the raw data in the body. Put the details in the appendix or leave them out of the paper altogether if they are not helpful to the reader.

Label each entry in your appendix with a capital letter—*A*, *B*, and so on—and a title, i.e., A. A Developed Site. List them in the table of contents also. Include only entries mentioned in the body of the paper. Pages are numbered as a continuation of the body.

Works Cited

The last entry in the appendix is the *Works Cited* section. In the MLA format, entries are alphabetized, indented five spaces for each line after the first, and double spaced. Figure 13.17 is the *Works Cited* page from the model research

4.0. CONCLUSION

Wind-generated power is feasible on the Southern Oregon Coast using a Nimbus WEC system. Though only rough approximations were presented, the estimated payback time of eight years makes the project cost-effective. If government money is received, the payback time would be greatly reduced. Once initial costs were paid, besides annual overhead (maintenance, security, etc.), all other revenue would go to profit for investors.

Estimating that the initial overhead of $149,280 would be paid in about eight years, twelve years of system work life would remain. In these twelve years at $18,821 per year, approximately $203,052 of new profit would be generated. (This figure includes $24,000 for maintenance and security.)

When people say that wind energy has come of age, they are, for the most part, correct. If an efficient system is put up on an adequate site, it can be a viable, cost-effective means of power production.

5.0. RECOMMENDATION

If the opportunity arises to invest in wind power, and the proposed project checks out in terms of location and system production, it should be a good investment.

Figure 13.16   Model Research Paper: Conclusion and Recommendation

paper. It, combined with the format discussed in Chapter 13, *The Research Paper*, is a handy model for many documentation situations you will confront in writing your own paper.

MISCELLANEOUS MATTERS

There are many conventions in research writing. The following are particularly important. For more information, see a complete guide to writing research papers.

1. When it is practical, introduce quotations and borrowed ideas by mentioning the author's name—Jones said, "   "
2. Use care to be completely accurate in what is recorded in quotation marks.

WORKS CITED

Atlas of Oregon. Director, William G. Loy. Eugene: University of Oregon, 1976.

Baker, Robert W. "Wind Power Potential of the Northwest Region." Power Engineering, 83 (June 1979):64–65.

Bartlett, Forest, Employee, Burns Concrete Company. Telephone Interview. Rexburg, Idaho, 26 March 1983.

Bingham, Edwin R. "Oregon." Colliers Encyclopedia. 1980 ed.

"Congress and the Nation." Ed. Martha V. Gottron. Washington D.C.: Congressional Quarterly, 1981.

Dahle, Leonard, President, Windmills International Diversified. Telephone Interview. Portland, Oregon, 27 March 1983.

Flavin, Christopher. "A Renaissance for Wind Power." Environment, 23 (Oct. 1981):39.

Garner, Gary, Engineer, Utah Power and Light. Personal Interview. Rexburg, Idaho, 25 March 1983.

Hamilton, Roger. "Can We Harness the Wind?" National Geographic, 148 (1975):825.

Lord, Gary. "Prospectus on the Numbus Wind Energy Conversion System." Portland: Windmills International Diversified, 1982.

"Noisy Windmill." Time, 2 June 1981, 79.

Rittenhouse, R. C. "Solar/Wind Power and Security." Power Engineering, 84.9 (1980):24.

Taylor, Ronald A. "Why Neighbors Squabble Over Progress." U.S. News and World Report, 2 Aug. 1982, 73–74.

Figure 13.17   Model Research Paper: Works Cited

If you need to include a comment of your own, place it in brackets [] within the quotation.

3. If you want to use only part of a quotation, indicate this with ellipsis marks (three spaced periods). At the beginning or middle of a sentence, indicate the missing words with. . . , leaving no space between the first dot and the last word. If you are leaving words out at the end of a sentence, add a fourth dot as a period to terminate the sentence.

4. Include quotations of four lines or less in the text. Indent longer quotations ten spaces on the left, and use double space (quotation marks are not needed in these cases).

5. Place a textual citation close to borrowed or quoted data. Do not try to make one citation fit an entire paragraph unless this accurately represents your use of the data.

6. Remember, direct and control the organization and flow of the document, establish the structure of the paper, and indicate it clearly as you proceed. Assimilate the data sufficiently so that you master the material and can write most of the paper in your own words with many of your own thoughts mixed with those from sources. Make no more than 10 percent of the body of the paper direct quotation; to have more would suggest that you have not fully assimilated the material.

7. Use the highest standards of typing and illustration. Write the master copy on bond paper and bind it with a durable cover. Reproduce illustrations in a variety of ways—through photographs, photocopyies, original illustrations—but not with pencil. Make the paper represent your best effort to be professional.

REVIEW QUESTIONS

1. What is the relationship of Chapter 11 to the writing of the research paper?
2. What is a *working bibliography?* What is its purpose?
3. What information is placed on a single bibliography card?
4. What are the bases of classification in most indexes such as the card catalogue and the *Reader's Guide to Periodical Literature?*
5. What is the greatest advantage of placing one bibliographical entry on each card?
6. How many ideas should you place on each note card? Why?
7. What are two good ways to correlate note and bibliography cards?
8. Why is note/bibliography card correlation important?
9. What is the purpose of the *2.0* on the sample note cards?
10. What are the basic differences among *summary, paraphrase, précis,* and *direct quotation?*
11. What is a *scientific attitude?*
12. What is *plagiarism?* Does it involve ideas as well as the words of others?
13. What is *documentation?*
14. What documentation system is used in the sample paper? What do the initials mean?
15. What is the typical content of the introduction to a research paper?

EXERCISES

1. Write a *Works Cited* page containing a minimum of ten entries. Include in your references books with single and multiple authors, weekly and monthly magazines, and journals with examples of separate and continuous pagination. Use MLA Style and alphabetize your entries. Use the guidelines and models discussed in this chapter as models.

2. Use the library to research at least two documentation styles other than MLA. Write a two-page paper comparing these systems with the MLA in terms of textual citation and bibliographical (*Works Cited*) format. Use the same references as examples for each style to show similarities and differences.

3. Using the college library, find the names of several journals related to your major field. Examine individual issues for explanations that identify the form of documentation used. Find the issue (often January) that explains the format or write for the style guide. In a short paper, identify the documentation style, cite examples of textual citation, and cite examples of the *Works Cited* section. If you find a copy of the style guide, save it for use in writing future papers in your major field.

4. Read an article in a periodical and write a descriptive abstract of it.

5. Read an article in a periodical and write an informative abstract of it. It may be interesting to use the same article for Exercises 4 and 5.

6. Chapter 12, *Writing a Proposal*, outlines the specifications for a 3000–4000 word (minimum ten double-spaced typewritten pages in body) paper. Before attempting to write a research paper, study the RFP carefully, review the present chapter, and then write.

14

Critique of a Research Paper

A *critique* is a review of a document to evaluate its overall effectivenss and accuracy. A carefully selected set of criteria is the basis of evaluation. In business and industry, you may need to evaluate a report, a product that your firm is considering for purchase, or possibly even one of your fellow employees who is up for a merit review.

A meaningful critique systematically applies known criteria, avoiding personal preference and bias. If the subject evaluated has weaknesses or strengths, the critic should cite them specifically. For example, it is not enough to say that someone has a weak writing style. Weak in what way? If the critic cannot objectify the criteria, strengths and/or weaknesses, pros and cons, and so on, then his or her opinion is not very valuable.

In this course, you may write a review of another student's research paper. Having just written a research paper yourself, you will be in a good position to evaluate the writing of another student to see how well it has met course writing standards.

Figure 14.1

This short chapter has three major sections. The first explains the format of the critique, the second establishes the criteria for evaluation, and the third provides two student models, including one critiquing the sample research paper presented in Chapter 13. The chapter concludes with review questions and exercises.

BASIC FORMAT

The critique follows the organizational pattern established throughout this book: a clear introduction, body, and conclusion. Even though you will be writing from a list of criteria, write your evaluation in the form of sentences and paragraphs.

In the introduction, along with the standard subject, purpose, scope, and plan of organization, include the name of the author and the title of the paper you are evaluating. In the body of your critique, provide specific support for both pro and con points. For example, it is not enough to say that the paper has too many grammatical errors; you should cite some examples as support for your criticism. If you disagree with the format or accuracy of the documentation, refer to specific notes and explain exactly what is wrong. In the conclusion to your critique, summarize your main points and, if requested by your instructor, assign a letter grade to the student research paper you have been evaluating.

It is important to remember that a critique can praise as well as find fault. If the author has done an excellent job in writing the introduction, say so. If the documentation seems especially accurate and thorough, say so.

One of the most important parts of your critique will be your evaluation of the documentation. To evaluate it accurately, look up the author's subject in the standard research sources such as the *Reader's Guide to Periodical Literature* to see if the author has used most of the primary sources available. In addition, examine four or five of the sources cited to see if the author handled documented material properly and accurately. Are direct quotations properly indicated? Are footnotes used where needed?

CRITERIA

Here is a list of criteria to use in evaluating a research paper:

1. *Topic.* Is the topic properly limited in scope for a paper of 3000–4000 words? Is it an interpretation-of-data paper?
2. *Research.* Has the author examined most of the primary and secondary sources available? Have the sources been studied and used properly? Has the author

properly cited research materials? Are the quotations used really meaningful, or are they simply space fillers? Has the author avoided plagiarism?

3. *Introduction.* Does the introduction follow the standard four-part introductory pattern? Is it an accurate introduction to what follows in the paper? Does the statement of plan of organization accurately introduce the main ideas or sections that follow? Is the research question stated succinctly and clearly?

4. *Body.* Does the author control the material, or is the author controlled by it? Does all data cited contribute to the central point—the research question? Has the author leaned so hard on one or two sources that the paper is simply an abbreviated, warmed-over version of them? Is the author's controlling influence evident in every section of the paper?

   Does each major section begin with an adequate introduction? Does the author provide meaningful transitions so that you never become lost or confused while reading the paper?

   Is the paper accurate technically? Do the data and the author's interpretation of it add up? Is the conclusion logical in terms of the data represented?

5. *Format and Mechanics.* Does the author follow the prescribed format cited in Chapter 12. Does the paper contain correct spelling, punctuation, grammar, word choice, and so on?

6. *Visual Support.* Has the author used meaningful drawings, maps, and photographs? Or are they mainly window dressing? Are they properly documented, and are they referred to in the paper?

7. *Conclusion.* Is the conclusion clearly stated? Does the evidence support it? Is it appropriate in length, detail, depth, and so on for this paper?

Although the criteria are listed in seven sections, it is not necessary to mention them in exactly the same order when you apply them in writing your critique.

MODELS OF STUDENT WRITING

The following models are clearly organized, cover most or all of the criteria, supply specific details on points criticized, and meet format requirements.

MODEL 1   **CRITIQUE**

This paper provides a complete evaluation of the research paper entitled *The Feasibility of Wind-Generated Power on the Southern Oregon Coast Using a*

Nimbus Wind Energy Conversion System, written by Clayton Dahle in fulfillment of the English XXX writing assignment. This critique will cover all of the basic criteria for evaluating research papers as has been outlined in class.

The good choice of topic was one of the major factors in making this a very worthwhile research paper. The topic is narrow with a specific point to prove, yet broad enough to include a variety of sources. By simply looking at the title, the reader may visualize must of what will be covered in the paper. It is also a topic that allows visual support and application of important principles such as definition, description of a mechanism, classification, and interpretation of data.

Although there was room for some improvement, the research was extensive enough to cover all aspects of the topic. Besides the numerous interviews necessary for this subject, the research was well rounded, including such sources as an atlas, encyclopedia, Congressional report, and several periodicals. The overall research was handled well, although several places, such as the beginning of the paragraph on "Blade Diameter" (p. 17), Appendix B and C, and the price per kilowatt figure used on p. 26, seemed to need some indication of source. More care in documentation would also have eliminated such mistakes as the bibliography entry on *U.S. News and World Report* in which the stated article appears on p. 46 instead of pages 73–74. Other than these small details, it is obvious the author put great effort into the research and deserves due credit.

Another major strong point of the paper was that the reader was not left to speculate on what was happening. The whole paper was well structured and provided good introductions and transitions for each subject change. In addition to several summaries, a complete well-written introduction and fact-filled conclusion let the reader know where the author was headed.

The overall mechanics and format of the paper added to its effectiveness. It was well organized, attractive, and easy to follow, with few detracting punctuation and spelling errors. This allowed the reader to concentrate on what was being said rather than on how it was said. The clarity of the paper was also greatly enhanced by the proper use of illustrations and calculations. Besides suggesting that Figure 1 face page 4 where it was mentioned (instead of with page 3), the mechanics and format were entirely effective.

The overall technical accuracy of the paper was professional, covering everything from the price of concrete to the cost-effectiveness of the entire project. After the numerous interviews and other research performed, it was obvious that the author had a sound understanding of the problem and a clear idea of the solution.

The good choice of topic, extensive research, and clear presentation made this paper a success. It is clear that the author went to great lengths to cover all the criteria for writing a good paper. A few small improvements would help but in comparison to the total work done these would be minor. The author has done an excellent job and the paper is worthy of an A− grade.

MODEL 2   **CRITIQUE**

The purpose of this paper is to review a research paper written by
_____, entitled *The Selection and Use of a Home Computer*.
Evaluation will focus on the topic choice and introduction, research, composition,
and conclusion. Following these, the summary will restate the major findings
and assign a letter grade.

Topic Choice.   Comparing computers was a good choice for a research
paper, and defining the procedure for finding one "best" computer provided
the necessary space and emphasis on interpretation of data. The title, however,
should have been narrowed more to imply research rather than report.

Introduction.   Though somewhat wordy, the introduction does an adequate
job, giving the subject, purpose, scope, and plan of organization. It stated
what was going to be addressed and explained the background of the problem.

Research.   Research for the paper was thin, limited to four periodicals
and a price list. Three of the sources were accurate and the fourth, from *Fortune*,
February 1982, does not exist. The data cited was adequately synthesized into
the work, which helped to support some of its statements. When using information from *U.S. News and World Report*, the author failed to note this. Throughout
the paper, the author expressed a satisfactory knowledge of the subject but
needed additional support to develop breadth and depth.

Writing.   The main thrust of the paper did have a clear pattern of progres-
sion. From beginning to end it was clear that five popular computers were
compared and the "best" was to be chosen from among them. There were
several places in the body where wordiness and ambiguous explanations clouded
progress. The explanation of the computer was good, but fuller explanation of
"byte" or "sprite graphics" would have helped this reader. A tighter writing
style and fewer punctuation errors would have made the technical language
easier to understand also.

Several times, deviations from the prescribed format appeared in the paper.
Instead of a well-defined three-level outline, the author used only headings
and subheadings. When the latter appeared in the paper, sometimes they were
in the middle of a sentence instead of flush with the margin. Also, only two
textual citations followed MLA Style, making it difficult to relate them to the
"works cited" list. A few illustrations would have made a better presentation.
The graph was good, but more pictures and charts would have opened up the
paper and provided needed support.

Conclusion.   The conclusion was short and to the point, finding the XXXXX
64 the best of the five tested. Some summary and explanation are needed to
make the ending less abrupt.

Summary.   The approach to the assignment was adequate, although the
writing was often weak. The documentation oversights, limited bibliography,
and lack of illustrations hurt the overall presentation. These weaknesses, plus
the fact that the paper was nearly 1,000 words below the 3,000 limit, make a
C grade seem appropriate.

REVIEW QUESTIONS

1. What is a *critique?*
2. What is the basic pattern of organization for a critique.
3. How can you support your criticisms, both pro and con?
4. How should you go about evaluating the documentation of the paper you are going to critique?
5. What are the seven points that you should cover in the critique?

EXERCISE

By now the writing assignment should be clear. You are to write a 500–750-word critique of a research paper given you in class. Use the criteria listed in this chapter as well as the RFP from Chapter 12 that established the original criteria for the research paper. Read the paper carefully for organization, clarity, and accuracy. Go to the library and check the more common sources—*Reader's Guide*, periodical indexes—to see if the author has covered most of the data available. Look up at least four of the sources to see if they have been handled correctly in the documentation. Check to see that the information has been properly noted in the text.

Once you have reviewed the paper and the documentation, write a clearly organized critical review, using three-part organization. Be sure your own paper is a model of clarity in organization. Support your criticisms (pro or con) with specific references and examples from the research paper.

PART V
EDITING TECHNICAL PROSE

15

Style in Technical Prose

Technical style wastes few words, stresses what is important, uses precise language, avoids double meaning, and is grammatically correct. It avoids the humor, informality, ambiguity, play on words, and so on, common to some writing. Technical prose should communicate information clearly, efficiently, and without much fuss!

This chapter discusses economy, emphasis, concreteness, clarity, and correctness. It is simple to improve your style: read the brief explanation, study the negative and positive examples, revise the sentences, and then discuss your work with your classmates and instructor. Sentences in the exercises are from college classes, business, and industry.

ECONOMY

Economy in writing is much the same as in automobiles: a car that goes forty miles on one gallon of gas is called *economical*; a writer who states something clearly but uses no extra words is economical too.

The following eight important rules lead to an economical style; learn to apply them properly, and you will cut out 25–40 percent of the words in your reports!

1. Combine sentences.
2. Cut repeated words and ideas.
3. Cut *who, which, that.*
4. Cut *there.*
5. Cut unnecessary prepositions.
6. Cut unnecessary nouns.
7. Cut empty verbs and verb phrases.
8. Prefer active voice.

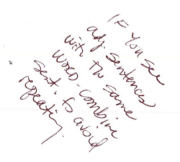

Combine Sentences

As you review a document, look for adjacent sentences that repeat words—nouns, pronouns, or verbs. A *noun* is a part of speech that names a person, animal, place, thing, or idea—*Bill, horse, New York, computer, love.* A *pronoun* is a part of speech that takes the place of a noun—*I, you, he, she, it, we, you, they.* A *verb* is a part of speech that shows action or state of being—action: *feed, close, move;* state of being: *is, am, are, was, were, be, being, been.*

Combining sentences ties ideas closely and produces economy by cutting unnecessary repetition of words and ideas. Study the example:

Original: The camshaft is enclosed in the engine block. It has eccentric lobes (cams) ground on it for each valve in the engine.

Revision: The camshaft enclosed in the engine block has eccentric lobes (cams) ground on it for each valve in the engine.

Camshaft and *it* repeat ideas in adjacent sentences. By combining the sentences, you tie information in the two sentences together and cut two words: *is* and *it.* This is a start towards economy.

Study another example:

Original: I will divide the object into three parts. First I will describe the inside of the pot. The second part will be the outer sides of the pot. The final part will be the outer base. (36 words)

Revision: I will divide the object (pot) into three parts: the inside, the outer sides, and the outer base. (17 words)

The original in the second example is grossly wordy. Notice all the repetitions: *part, will, pot.* The revision cuts 19 words!

Here are a few words of caution. If the combination of two sentences is over twenty-five words, it may not be wise to combine them. If the combination is difficult to understand, definitely do not combine them.

Follow these guidelines:

1. Avoid combining sentences with *and*. If you see no other way to do it, keep the short sentences.
2. Combine sentences two at a time. More may confuse you.
3. Do not cut necessary information.

EXERCISE

Directions: Underline repeated words or ideas. Then, combine sentences to eliminate unnecessary repetitions and gain economy. Evaluate the results for each sentence: are the changes worthwhile?

1. The Sureshot 9422 magnum, is a twenty-two caliber rifle with lever action. This rifle has a detachable clip which holds fifteen shells.
2. The Sureshot 122 semi-automatic rifle has a revolutionary rotary magazine. The rotary magazine is a small device that holds ten shells and is placed directly under the chamber.
3. The atmosphere of Bowen's is elegant. It is done in heavy dark wood beams and mirrors.
4. This paper will deal with the three methods specifically listed in the 1954 Internal Revenue code. They are the straight-line method, the declining-balance method, and the sum-of-the-years-digits method.
5. To learn which car best meets the needs of this elderly couple, each of

these cars will be matched up with ~~these~~ five characteristics. ~~The five most~~ important characteristics, ~~in a car for this couple are~~ fuel economy, comfort, price, trunk space, and acceleration. (Cut to about 26 words).

6. The view camera, ~~is~~ the oldest and largest type of camera available, ~~This camera uses~~ a "through-the-lens" focusing system.

7. The findings for this report came from a study made by a team of researchers from Idaho State University, ~~The study~~ is entitled "Alternative Systems for Solid Waste Disposal in a Three-Country Region."

8. The purpose of this paper is to describe an unknown object. ~~The object is~~ small enough to fit in the palm of a hand ~~and weighs~~ approximately one ounce.

9. Each cell has a unique mnemonic or technical name. ~~This technical name~~ is important in both design and layout.

10. One way to achieve the required alignment is through the use of ~~an end latch. One such centering device is~~ located at each end of the span.

Cut Repeated Words and Ideas

The previous section explains how to cut repeated words and ideas by combining sentences. This section shows you how to cut unnecessary repetition within a single sentence.

REPETITION OF WORDS

Study the following example:

Original: The cell* catalogue not only includes the usual gate and flip-flop cells, but more complex shift, register, counter, and latch cell functions as well. (24 words)

Revision: The cell catalogue not only includes the usual gates and flip-flops, but more complex shift, register, counter and latch functions as well. (22 words)

* Cells are the micron-level components of integrated circuits (IC) or *chips* that make up the memory of a computer.

The revision removes two unnecessary repetitions of *cell*.

REPETITION OF IDEAS

Sometimes wordiness results from repetition of ideas.

Original: Is it possible at this point in time to limit our liability for any damage to the work itself? (19 words)

Revision: Is it possible at this time to limit our liability for any damage to the work itself? (17 words)

The revision cuts two words—*point* and *in*—in the first line. *At this point* and *at this time* say the same thing.

Repetition of ideas also weakens this example:

Original: The hull of the yacht was red in color, rather large in size, and certainly graceful in shape. (18 words)

Revision: The yacht's red hull was large and graceful. (8 words)

It is insulting to tell the reader that *red* is a color, that *large* is a size, and that *graceful* relates to shape.

Avoid phrases such as these:

POOR	BETTER
large in size	large
red in color	red
graceful in shape	graceful
final conclusion	conclusion
close proximity	near
any and all	any
return back	return
advanced forward	advanced
25 in number	25
apologize and say you're sorry	apologize
in this day and age	today

EXERCISE

Directions: Cut unnecessary repetition of words and ideas.

1. The Standard Cell data sheets provide the information you need to perform circuit design on your circuit.
2. The core cells are those cells placed in the interior of the circuit that causes the desired logic function to be realized.
3. *Interconnect* describes the means by which the cells are linked together to create the logic function.
4. Free gifts will be given at each year's annual credit union banquet.
5. The only way we can attract and keep good people in the position is to be candid about the demands of the position.
6. The topic of an efficient solar greenhouse is a good choice for a topic.

7. The conflict in Zimbabwe, ~~probably~~ stemmed from many incidents, but this research paper ~~will try to~~ prove that the land was the main cause ~~of the conflict in Zimbabwe.~~

8. A bibliography on ~~the subject matter of~~ mine drainage control will be prepared.

9. It is proposed to chart out a detailed program plan ~~in consultation~~ with the group leaders of the four disciplines ~~pooled~~ together for this project.

10. The clinker from each cooler is conveyed a distance of about 240′ by a dual drag chain conveyor, 20″ wide.

Cut *Who, Which,* and *That*

The relative pronouns *who, which,* and *that* are often deadwood—words that can be deleted without changing the meaning of a sentence. An easy way to estimate their value is to read the sentence without them; if it still makes sense, delete them. Study this example:

Original: The equipment which is located in Boise is all surplus.

Revision: The equipment located in Boise is all surplus.

Although *which* and *is* were deleted, the sentence still makes good sense. In fact, it is now more economical and emphatic. The *be* verb—*is, am, are, was, were, be, being,* and *been*—often accompanies the relative pronouns. If the verb is deadwood, strike it also.

Sometimes only the relative pronoun is deadwood:

Original: The general subject that I have chosen to research is the negative and positive aspects if the Cambridge Diet.

Revision: The general subject I have chosen to. . . .

Now see what you can do with the exercise. Remember, if deleting the relative pronoun and verb makes the sentence awkward or difficult to follow, do not delete them.

EXERCISE

Directions: Cut the relative pronouns *who, which,* and *that* and accompanying verbs and wordiness—when they are deadwood.

1. The characteristics that are used in evaluating these weapons are weight, length, rate, and type of magazine.

2. The three weapons that will be dealt with in this report will be the United States M3A1, the Swedish 9mm Gustaf, and the Israeli 9mm UZI.

3. The first characteristic ~~to be~~ discussed is ~~the~~ weapon weight, ~~which is~~ an important factor in choosing an assault weapon.

4. The M3A1 has a firing rate of 450 rpm (~~rounds per minute~~), ~~which is~~ the lowest of ~~the~~ four.

5. Each dress will be judged according to the four requirements ~~that I have~~ previously mentioned.

6. ~~K-Mart is the store that I went to to~~ find a dress ~~that I might buy~~.

7. The following ~~is the~~ format ~~which will be utilized~~ *will be used* for the Power Group Monthly Engineering Report.

8. Most instrument mounting devices ~~which were~~ fabricated by the instrument department follow standard details. *we believe*

9. After conversations with N.S.D. staff, ~~it is our belief that the~~ *we believe* greater volumes and pressures will be beneficial to the artificial lift project.

10. All information indicates that the Garrett (3rd generation) is performing ~~in a successful manner~~ *successfully*.

Cut *There*

There and *it*, usually called *expletives*, take the position of the subject in a sentence. Expletives usually appear with the verb *to be* and sometimes with a *who, which,* or *that*. If the expletive fails to add emphasis, variety, or economy, cut it.

Study this example:

Original: There are several unique characteristics of a memory typewriter which are not available with the electronics. (16 words)

Revision: Several unique characteristics of a memory typewriter are not available with the electronics. (13 words)

The revision deletes *There are . . . which . . .* as deadwood, leaving the remainder of the sentence unchanged. The revision is economical and emphatic. It emphasizes *characteristics* by moving this key term toward the beginning of the sentence.

Sometimes you can remove only the expletive, sometimes both the expletive and the verb. Study this example:

Original: There appear to be several factors supporting the change in price. (11 words)

Revision: Several factors appear to support the change in price. (9 words)

The revision is economical and emphatic. *Factors*, buried in the middle of the original, is now in an emphatic position at the beginning of the sentence.

If used sparingly, expletives provide variation from the usual subject-verb-object order of most sentences.

EXERCISE

Directions: Cut *there*, *it*, and accompanying wordiness.

1. ~~There is often a misunderstanding by the public on~~ what a sports car really is and is expected to do. ~~Often~~ The public *misunderstood* what

2. ~~There are~~ two primary considerations in this *the* criteria: the friction potential and aerodynamic design.

3. ~~There are~~ 345 pieces of equipment in the Kananga area.

4. ~~There are~~ a lot of speakers ~~that~~ could be evaluated, but ~~for the purposes of this paper~~ I have selected three.

5. ~~It is~~ believed ~~by~~ many that sugar and saccharin are no better or worse for a person's health than any other food. *to be reviewed*

6. ~~There are~~ two basic considerations ~~which must be reviewed~~ when determining bus widths: current density and voltage drop.

7. With the increasing number of women working outside the home, ~~there is~~ definite ~~need in most~~ homes for food which is quick and easy to prepare and serve. *most homes need*

8. ~~There are~~ three basic methods for feeding paper into a printer: *are*, roller feed, tractor feed, and a combination of both.

9. ~~It is our belief that the~~ Existing dryer, cyclones, and scrubber are adequate.

10. ~~It was found that~~ the preliminary design and engineering for the proposed plant meet the capital cost estimate.

Cut Unnecessary Prepositions

Prepositions, words like *in, on, over, under, for, at, by,* and the phrases formed by them, can clog writing just as cholesterol clogs arteries. If sentences have more that two or three prepositional phrases (*on the floor, behind the house, and so forth*), the sentence probably needs editing.

Original: In the areas of technical writing and of business writing for college classes, most of the textbooks available are out of touch with the reality of the work place. (Count them; there are 8 prepositional phrases in this sentence or 29 words.)

Revision: In business and technical writing, most college textbooks don't touch the reality of the work place. (2 prepositional phrases or 16 words)

The revision is vigorous, economical, clear, and emphatic. The string of four prepositional phrases at the beginning of the original is particularly deadly.

Some writers also habitually use prepositional-based constructions, sometimes called *prepositional idioms*, which are also wordy. Usually these forms can be reduced to a simple, economical form.

NEGATIVE EXAMPLES	BETTER FORMS
with the purpose of	to
for the purpose of	so
with regard to	regarding
next to	by
in close proximity to	near
not far from	near
in the region of	around
in the area of	in

Prefer the short form and gain economy in your writing.

Although prepositions can clog a sentence, they are a necessary part of our language. Your goal should be to rid your sentences of excess prepositions.

EXERCISE

Cut unnecessary prepositions and preposition-based clichés.

1. In ~~the area of~~ engines, Suzuki has done the most engineering.
2. She will ~~bring the situation to the attention of Dawn Baird~~, the ~~Senior Project Secretary.~~ *notify* Dawn Baird.
3. ~~In the case of~~ an individual charging time to several projects, ~~he~~ will submit ~~his~~ time sheet to the project engineer.
4. ~~The need for~~ clear and precise work instructions to performing organizations is essential to assure ~~correct~~ completion of required work in accordance with ~~established~~ requirements.
5. Thank you for ~~the acceptance of~~ the initial memo.
6. My two-year degree includes knowledge ~~of the languages~~ of COBAL and Fortran.
7. I would appreciate an ~~opportunity to be~~ interviewed *interview* ~~by you.~~
8. I have ~~just~~ finished one year out of a two year program at Ricks College in ~~the department of~~ electronics engineering technology.
9. All four locomotives have been fitted with ~~these~~ special elements and will be closely monitored for the first few change intervals ~~in order~~ to determine the life of the elements.
10. In ~~the area of~~ equipment needs, ~~we feel~~ it is important that we make our needs known before ~~other needs in our group and in the company in general limit~~ our opportunities to status quo or even to reductions.
are limited

Cut Unnecessary Nouns

A *Noun* is a part of speech that names a person, animal, place, thing, or idea. Some are *empty*—a form of wordiness that unnecessarily classifies the subject. Study this example:

Original: The loss of the craft is an unfortunate circumstance. (9 words)

Revision: The loss of the craft is unfortunate. (7 words)

The revision deletes *circumstance*, the empty noun, and the article *an*. *Articles* are the little words *a, and,* and *the*. Because readers recognize that the *loss of the craft* is a circumstance, there is no reason to use the word. The revised sentence is economical and emphatic.

Empty nouns generally occur in the following conditions:

1. after the verb be-*is, am, are, was, were*.
2. after an introductory *a, and,* or *the*.
3. after a restrictive adjective, such as *unfortunate* in the first example.

An *Adjective* is a part of speech that modifies or limits a noun (a *tall* boy). Other empty nouns include the following:

- a difficult condition
- a critical nature
- an important role
- a memorable matter
- a difficult circumstance

Study one more example before you turn to the exercise:

Original: Buying hi-fi speakers can be a very time-consuming and exhausting experience.

Revision: Buying hi-fi speakers can be time-consuming and exhausting.

The original has all the basic characteristics of an empty noun: preceded by 1) the verb *be*, 2) an article *a*, and 3) adjectives *time-consuming* and *exhausting*. The revision cuts the empty noun *experience* and the article *a*. Remember, although some empty nouns may have only one or two of the basic characteristics, you should still delete them.

Cutting empty nouns is one more way to gain economy in your writing. Now turn to the exercise.

EXERCISE _____

Directions: Cut empty nouns and associated wordiness.

1. The data was found in the August 1981 ~~issue~~ of *Stereo Review.*
2. Sailing ~~is an activity that~~ requires a keen sense of wind direction.
3. The question of proper business ethics ~~in this situation~~ has always been ~~an~~ intriguing ~~one~~ for my partner.
4. Coal removal equipment ~~has the role of~~ removing each seam ~~as~~ exposed by either the shovels or the draglines.
5. Employing the dragline-truck-shovel approach is ~~a matter of some~~ importance because of the cost savings.
6. This concept ~~appears to be the approach which~~ combines the cost effectiveness of dragline stripping with the mobility of truck/shovel operation.
7. Anticipating pre-development mining and a gradual build up is ~~an~~ important ~~element~~ in the conceptual mine plan for the recovery of 60 million tons.
8. Traveling the escapeway every six weeks is an important procedure to guarantee mine safety.
9. Bad weather ~~conditions~~ prevented our takeoff.
10. This ~~is a matter~~ concerning morale.

Cut Weak Verbs and Verb Phrases

A strong verb makes a sentence concrete. A *Verb* is a part of speech that expresses action (walk, jump, open, close) or being (is, am, are, was, were, be, being, and been.)

ACTION VERBS

A strong verb is specific. Study the following example:

Original: The climber *moved* into a narrow chimney in the rock face.

Revision: The climber *edged* into a narrow chimney in the rock face.

Move in the original sentence is weak because it doesn't tell exactly what the climber is doing. In the revised sentence *edged* (struggled, slipped, or crawled) is specific and, therefore, stronger.

Sometimes you can make a verb phrase more specific by reducing it to a single word. A *Verb Phrase* is a word or group of words that express an action performed by the subject or a state of being (The man *was sent* home at 9:30.) Consider this example:

Original: Two systems *are available to verify* the resources used up, compared to the resources acquired during a given period. (19 words)

Revision: Two systems *verify* the resources used up, compared to the resources acquired during a given period. (16 words)

The revision reduces the verb phrase *are available to verify* to a single strong verb, *verify*. Choose strong verbs to eliminate wordiness.

Being Verbs

Being verbs—*is, am, are, was, were, be, being, been*—are often wordy and imprecise. You can often delete them as deadwood or substitute a strong, precise verb in their place. In the following example, all forms of *being* verbs are italicized:

Original: To eliminate the confusion about when, where, and what escapeways *are* to *be* traveled, the following criteria will *be* followed to insure compliance with this regulation. (26 words)

Revision: To eliminate confusion about when, where, and what escapeways to travel, use the following criteria to insure compliance with this regulation. (21 words)

The revision cuts the weak verbs *are* and *to be* in the second line, and *be* and associated wordiness in the third line. These are worthwhile changes; however, a word of caution is in order. Although *be* verbs are often weak, they are an important part of the language. Your goal is not to remove *all;* but to cut weak usage.

To eliminate weak verbs in your writing, underline all verbs in several pages of your own writing and analyze your choices. If they are general or weak, substitute strong verbs. Repeat this procedure every few weeks.

EXERCISE

Directions: Cut or replace weak verbs and phrases in the following sentences.

1. Information ~~to be~~ used in this paper ~~was obtained~~ *came* from an article by David Ranada.
2. The first standard ~~to be~~ considered is the price.
3. Before handing in the timesheets, ~~it will be the responsibility of each~~ individual~~, to obtain~~ *will get* *must obtain* the ~~required~~ approval of the project engineer.
4. Anyone requiring office or engineering supplies will ~~make his needs known to~~ *inform* Jo.
5. ~~I feel~~ the best policy *is* to adopt a price schedule ~~which would be~~ comparable to that of commercial shops in the area.
6. ~~If any~~ *B*ranches ~~are found~~ with excessive current density, ~~they~~ should be widened.

7. The collision made a dent in the housing but didn't affect the operation of the equipment.
8. The lip is attached to the body just below the rim.
9. He made an attempt at finishing the job, but it was clearly half-hearted.
10. Aeroquip hose has been found to be too fragile for the environment and was replaced with double-strength steel piping.

Prefer Active Voice

Many business and technical writers frequently use the *passive* verb form. However, most books on effective writing style advise writers to prefer the *active* form. To be in control of your own writing, you must understand differences between the two forms and know when to use them. This section teaches you to recognize active and passive verbs, explains when to use them, and then gives you practice in revising and creating them.

The following examples and analysis will help you understand differences between sentences containing active and passive verbs:

Active: The report emphasized the lake's importance. (6 words)

Passive: The lake's importance was emphasized by the report. (8 words)

For a breakdown of basic sentence elements, study the following styles:

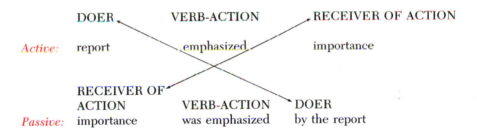

This analysis shows the main differences between active and passive sentences:

1. Active stresses the *doer* or subject performing an action.
2. Passive stresses *receiver of action* or result of action.
3. Passive requires additional words—for example, *was* and *by*—making it less economical.

The third difference holds the keys to recognizing the passive: there *must* be a form of the verb *be*, plus there *must* be at least one main verb called a past participle. If both parts are not present, a verb form is not passive—it is active! Here is a breakdown of the characteristic form:

BE VERB	+	MAIN VERB
is		Past Participle
am		—often ends in -en or -ed
are		—never ends in -ing
was	Plus	—some end irregularly
were		—gone
be		—flown
being		—sung
been		—etc.

Besides this characteristic verb form, many sentences containing the passive also end in a prepositional phrase that contains the *doer* or performer of the action. Study this example.

Passive: The door was closed (by a technician).

By a technician is a prepositional phrase explaining who closed the door. If *who* did it is not important, deleting the prepositional phrase results in a short sentence that stresses *what* was done:

Passive: The door was closed.

Did you spot the passive verb? *Was* is a form of the verb *be* and *closed* is the *main verb,* in its past participle form.

Here are a few more examples to help you solidify your ability to recognize active and passive verbs:

1. The weather has *been* cold for November.

Although *been* is a *be* verb, no main verb in the past participle is present; therefore, this sentence is active.

2. The report had *been submitted* three weeks earlier.

Been is a *be* verb; *submitted* is a main verb with the common *-ed* ending; therefore, this sentence *must* be passive. Now try your knowledge on a recognition exercise.

EXERCISE

Directions: Underline the verbs. If they are active, place *A* in the blank; if passive, place *P* in the blank.

   P   1. The lost report was found in a desk drawer.
   A   2. The group leader will be arriving next week.

↳ main verb end in ing

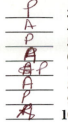

P   3. The project was completed with little difficulty.
A   4. The data are extremely significant.
P   5. The technician was ordered to a new post, pending investigation.
A   6. The operator shut the plant down according to standard procedure.
A-P   7. The data is being sent to Dallas by the chief engineer.
A   8. The chief engineer is sending the data to Dallas.
P   9. The equipment was destroyed in the hurricane.
A   10. The sequences of variations are the result of improper calibration.

The purpose of the discussion so far has been to acquaint you with active and passive verbs, and give you a surefire method for identifying them. Now it is time to consider some of the issues and problems related to usage.

Some editors and supervisors prefer the passive because they feel the active somehow destroys scientific objectivity by focusing on the *doer* rather than on *what* is done. The numerous statements that include *we did this* . . . or *I chose this material* . . . and similar references to the *doer*, they feel, detract from the information contained in the writing.

By contrast, many writers feel the passive is wordy and indirect. They feel it creates a world in which things happen but no one takes responsibility because the *doer* simply disappears.

Both points of view have merit. For a student writer, the best policy is to be fully aware of the limitations and benefits of each form—and then use the best form for a given situation.

LIMITATIONS OF THE PASSIVE

Sometimes the passive produces strained word order. Consider the following example:

Example Passive: It was suggested by Mr. Durn of the D. M. Weather Company that a steam stripper be installed. (18 words)

Revision/Active: Mr. Durn of the D. M. Weather Company suggested installing a steam stripper. (13 words)

The revision reads well, avoids the strained passive construction, and saves six words. Because the *doer*—Mr. Durn—is mentioned in the passive, it is easy to make him the subject of the sentence. Here the active is the preferred form.

Some engineers and business writers use the passive so extensively that they fail to mention the *doer* even in situations where the audience needs to know. In the following sentence, for example, some supervisors will want to know *who* failed to lubricate the bearing.

Example Passive: The bearing failed because it had not been lubricated on schedule.

Of course, if you do not want to identify the *doer*, the passive is a handy form!

Another problem of the passive involves incorrect use of cognitive verbs such as *learned, concluded, thought,* and *deduced*. A *cognitive verb* involves an action that logically could be performed only by a person. Some writers, in their dependence on the passive, defy logic, particularly if the actor is not mentioned. Review this example:

Example Passive:	In talking to some of the plants where a stripper* had been installed, it was found that the results have not been as expected. (24 words)
Revision/Active:	Several plants which installed strippers have had unexpected results. (9 words)

* Paint stripper

Obviously the passive, if used without some forethought, can hinder communication. Yet technical writers need both active and passive forms to meet a variety of needs. The next section explains when to use each.

WHEN TO USE THE ACTIVE

1. To emphasize actor or agent

   J. T. McRoy of Reece Research Institute obtained significantly better IGSCC detection by. . . .

2. To define responsibility

   Not: The instruments were tested one week before the damage occurred.

   But: Our own technicians tested the instruments one week before the damage occurred.

3. To use cognitive verbs logically

   Not: Based upon the above results,* it was assumed that the free NH_3 resulted from high pH in the neutralizer and scrubber.

   * Dangling modifier.

   But: Based upon the above results, our staff assumed the free NH_3 resulted from high pH in the neutralizer and scrubber.

4. To write instructions

   Not: The door to the control room will be closed in the event of an emergency.

   But: In the event of an emergency, close the door to the emergency room.

Use the active whenever the agent is at least as important as the result of the action.

WHEN TO USE THE PASSIVE

1. When the actor is unknown or unimportant, and a natural subject is not available
2. When emphasis, variety, and tone demand it

The passive is used appropriately in the following examples:

> Of the twenty-two state-listed species, the following are found on the IPGA: Grizzley bear, Northern Rocky Mountain wolf, . . .

Who found these animals is unimportant to the author (and reader) of this environmental impact statement.

> The unit was shut down due to an increase in the grid ΔP.

Here, again, the actor is probably of little consequence. The *what*, not the *who*, is what the sentence is about.

The following example, taken from a long report dealing with the history of intergranular stress corrosion cracking, again shows that *what*, not *who*, is of particular interest to the reader.

> A brief history of cracking in austenitic stainless steel piping has been presented in References 1, 2, and 3 and other documents. As early as 1965, small hairline cracks were detected in austenitic stainless piping in BWR* facilities. . . . Cracks were located only in the weld heat affected zone.
>
> * Boiling Water Reactor

To help you master the active and passive, the following exercises give you an opportunity to revise sentences of both forms.

EXERCISE

Directions: Underline the passive verbs and then rewrite the sentence in the active voice. If an actor is not supplied in the sentence, use a general term such as *staff*, *engineering*, or *technicians*.

1. At that time a thorough literature review was made which revealed that watered sludge could be disposed of.
2. The new ¾″ grid plate is presently being drilled at Paul Roberts' Machine Shop and will be delivered on January 18.
3. Most waterfowl on the IPGA during the hunting season are produced outside the area. (Are waterfowl manufactured?)
4. The preliminary design and engineering presented in the previous sections is used as the basis for developing capital and operating cost estimates.

5. The estimate for the Fluidized Bed Combustor facility was made utilizing details obtained from the preliminary engineering reports.
6. At least one site in each region will be visited by the inspection team.
7. A listing and description of possible technical gaps and research needs will also be prepared.
8. An investigation was made of filter media and filter design and two firms were found which could provide both the required paper and the right configuration.
9. Considering the high temperatures found in the first stage, it was decided to shut down the separator until operating limitations could be met.
10. The site selection was made by J. P. Jones from Engineering.

EXERCISE

Directions: To sharpen your skill in recognizing and using active and passive verbs, change the following active sentences into the passive. If it seems appropriate, leave out the actor—if one is included in the original.

1. Engineering is now examining design approach alternatives for recommendation to high-level engineering management.
2. We find no parallel source development funding budgeted.
3. Engineering determined through tests that a fluidized bed combustor can adequately burn and control refuse.
4. We have received your letter regarding J. M. Lowery as a candidate for the accounting position.
5. After talking with the shift supervisor, we have decided to change the head.   (Don't leave a dangling modifier at the beginning of your new sentence)
6. We decided to make no change at this time.
7. Backoff shot service disengaged the drill pipe from the coring assembly.
8. I found the Number 3 grade acceptable for strength and flexibility.
9. The Bronco II leads the trio in horsepower ratio at 15–1.
10. The critic considered weight and economy as important factors in the comparison.

EMPHASIS

As you write technical documents, your first priority is usually to get your ideas down on paper. Once this is accomplished, you can revise individual sentences to emphasize key ideas by altering word order in two important ways: stressing beginning and ending positions and using *loose* and *periodic* sentences.

Beginnings and Endings

A sentence has three positions: a beginning, middle, and end. The end is the most emphatic, the beginning a close second, and the middle the least. Data placed at the end or beginning of a sentence has more impact than data buried in the middle. Study the following example:

Original: There are <u>eight pieces of equipment</u> available at the job site.

Revision: <u>Eight pieces of equipment</u> are available at the job site.

The revision emphasizes *eight pieces of equipment* by placing it at the beginning of the sentence. Further revision can shift this same information to the end of the sentence:

Revision: Available at the job site are <u>eight pieces of equipment</u>.

Either version emphasizes key data more than the original.

Beginnings and endings also relate to the order of ideas in a series. A *series* is a listing of grammatically equal items—noun, noun, and noun—separated by commas. Compare versions of the following sentence:

Original: The violent winter storm killed twenty-five, caused hundreds of accidents, and stranded many.

Revision: The violent winter storm stranded many, caused hundreds of accidents, and killed twenty-five.

The revision has greater emphasis because it places *killed twenty-five*, the most significant fact, in an emphatic position.

EXERCISE

Directions: Change emphasis by placing the key idea at the end of the sentence.

1. Mr. Jones seems a good solid engineer with the proper mix of design and project responsibility. (Stress *good solid engineer*)
2. They will replace the bad cable at no cost to Acme. (Stress *the bad cable*)
3. Washington, D.C., home of the world famous Smithsonian Institute of Technology, is located on the Potomac River. (Stress *Smithsonian Institute of Technology*, then stress *Washington, D.C.*)
4. Many presidents consider fund raising to be the least desirable role of the office. (Stress *fund raising*)
5. This design of this hull makes it totally unstable in extreme weather conditions. (Stress *totally unstable*)

6. Your visit to Washington includes a meeting with President Reagan, dinner in a local restaurant, a one-hour cruise of the Potomac, and a two-hour rest period. (Revise the series statement to stress the key experience)

Loose and Periodic Sentences

Closely related to the discussion on word placement are *loose* and *periodic* sentences. A *loose sentence* identifies the subject and verb in the first few words. This pattern is by far the most commonly used in the language. As a grammatical term, then, *loose* does not mean that a sentence is inferior in any way. Study a few examples:

> *Loose Sentence:* The inspectors will visit at least one site in each region.

> *Loose Sentence:* Plants were established in four sections of the nation.

The *periodic sentence* differs from the loose by delaying some key element, usually the subject or verb or both, until the end of the sentence. To clarify your understanding of the two terms, study the following sentences:

> *Loose:* Here are some ideas concerning the proposed workshop.

> *Periodic:* Concerning the proposed workshop, here are some ideas.

> *Loose:* The students studied when all else failed.

> *Periodic:* When all else failed, the students studied.

As you can see, changing loose sentences to periodic involves mainly a change in emphasis—putting a key idea at the end of a sentence. Such changes, if used only occasionally, are an excellent way to emphasize a key idea.

A TOUCH OF PERIODICITY

If you find it difficult to move key ideas to the end of a sentence, you can gain some variety in your style by delaying the subject and verb with occasional introductory clauses and phrases—just a "touch" of periodicity. Consider these examples:

> *Original:* Acme's computer will be used advantageously if there is need to store and analyze field data.

> *Revision:* During the study, Acme's computer will be used advantageously if there is need to store and analyze field data.

The simple addition of *during the study* ads variety and, of course, a touch of periodicity. You could easily make this sentence even more periodic by also moving the *if* clause to the beginning of the sentence.

      Revision: During the study, if there is need to store and analyze field data, Acme's computer will be used advantageously.

The loose sentence should continue to be your most common pattern; however, for variety and for emphasis the periodic sentence and/or mildly periodic sentence have much to recommend them.

EXERCISE

Directions: Revise the following sentences to give them variety and periodicity.

1. Enter the applicants' names in the log as résumés and/or applications are received.
2. It must be sent to ARS if no "Document" stamp appears on the résumé.
3. The following action should be taken if the recruiter indicates the résumé is to be routed to departmental employees.
4. One problem—parts availability—was discussed with Jeffrey.
5. Maintenance costs and down time tell us we are spending many more work hours on the wide heads than the narrow heads.
6. The cost of the overhaul will run about $220,000 as near as I can figure.
7. The Bureau *shall* assume the entire responsibility and indemnify Deepseas Inc. for all damage or injury of any kind to all personnel in China, whether employees of the Bureau or otherwise.
8. We are also using this opportunity to inform you that we have been able to fill the position of Assistant Process Facilities Manager from another source.
9. No parallel source development funding is available based on information received to date.
10. This includes costs for all engineering, construction, indirects, fees, sales tax, and contingency.

CONCRETENESS

Writing may be concrete, abstract, or some place in between. *Concrete* words—nouns, verbs, adjectives—relate to the senses of sight, sound, taste, touch, and smell. If you can see something, it is concrete; if you can hear it, it has a frequency and range—concrete characteristics; if you can taste, touch, or smell it, it is concrete.

    Of course, some words are more concrete than others. The following example illustrates these levels:

Abstract———————————————→Concrete
device—tool—hand tool—hammer—claw hammer—etc.

Level

1. *Device* is completely broad, potentially includes everything ever invented.
2. *Tool* is more restrictive but still very broad.
3. *Handtool* is much more restrictive but still somewhat broad.
4. *Hammer* is specific enough for most documents but can still be made more specific.
5. *Claw hammer* identifies a specific tool used for driving and pulling nails.

Claw hammer can be made even more concrete by identifying its weight, manufacturer, and specialized design. The level appropriate in a document depends upon its purpose and audience. In general, good writing is on levels three–five. For further insight compare the following examples:

Original: Construction is up in the state for the last reporting period. Although significant regional variations exist, the general trend is positive. If it continues at this pace, employment in the industry is certain to rebound from the frustrating lows recently experienced.

Revision: Housing starts are up in (state) in the second quarter of 1985. Although several areas report little change, the overall trend shows an increase of 15 percent or 10,000 starts. If this trend continues, employment in the basic building trades will rebound from the frustrating lows of 35 percent unemployment recently experienced.

The original vaguely refers to problems but never provides any concrete information. The revision, on the other hand, identifies the following:

- state (XXXXXX)
- quarter (second)
- year (1985)
- size of increase (15 percent)
- number of house starts (10,000)
- who will benefit (basic building trades)
- how low unemployment has been (35 percent)

The original is unacceptably general; the revision is concrete and informative for most readers. A good rule to follow is the following: *Be concrete to the level that meets the needs of your readers.*

Abstract writing refers to ideas, opinions, and feelings: a democracy, supply-side economics, good, bad, normal, low, high, sad, happy. Technical writers use these terms to state an opinion, theory, judgment, interpretation, conclusion, and bias.

In a work situation, for example, you may be asked to evaluate a piece of equipment. You will probably write statements such as the following:

Example: —The construction is poor.
              —Location of the controls is poor.
              —Documentation is inadequate for our purposes.
              —The secondary shaft turns too fast.

If you have been asked to provide an informal evaluation, these statements are probably sufficient. If precision is requested, personal opinions need substance to back them up.

Questions: —What does *poor* mean? What is poor about it?
              —What is *poor* about control location?
              —What is *inadequate* about the documentation?
              —How fast is *too fast?*

Revision: The construction of the machine is poor—welds have not been filed, support seems inadequate for the weight of the machine. The controls are poorly located—the operator has to reach across the cutting surface to activate the oiler. The operator's manual is inadequate with few illustrations and no explanation of how to set up and use the milling attachment. The secondary shaft runs at least 200 rpms more than in other similar machines.

The revision communicates more effectively because it makes the abstract concrete by providing supportive detail.

Basic Rules

Here are four basic rules to help you be more concrete in your own writing:

1. Replace vague general words with specific ones.

   Original: One building was destroyed in the accident.

   Revision: One *wing* of *Old Main* was destroyed in the *fire*.

2. Support important generalizations.

   The fall 1983 earthquake was cataclysmic in Challis.

   (If the writer does not go ahead and tell us what *cataclysmic* means to him, readers will have different ideas based upon their own experience.)

   Rule: *Generalizations fail if they are not supported by concrete detail.*

3. Make the unfamiliar familiar.

   Original: The *boom vang* is useful to control sail shape when sailing into and down wind.

   (What is a *boom vang?* the reader asks.)

Revision: The boom vang, *a system of rope and pulleys hooked between the base of the mast and the middle of the boom,* is useful in controlling sail shape when sailing into and down wind.

(This version helps the reader to understand by explaining a term used in sailing jargon.)

Using an analogy, a comparison between the unknown and something the reader knows, is another good way to make your writing specific.

4. Support important abstract terms with concrete detail.

Original: The results were relatively inconclusive.

Revision: The results were inconclusive—*only three of ten were positive.*

Original: The product was very hard.

Revision: The product was hard, *approximately 8 on Mohs' scale.*

Notice how the authors of the following sentences made their statements specific by adding data after their opening generalizations:

Positive Examples

A set of parameters is established—*how many berths, price range, bio sketch of proposed owners, including their previous boats.*

Sometimes the difference between good and mediocre performance is all those fine adjustments—*Backstay tension, traveler position, sheet leads, sail cut.*

EXERCISE

Directions: Underline unsupported generalizations in the following sentences and make them more specific.

1. The temperature was too far above the 600-degree cutoff.
2. Production was off significantly in last year.
3. Farmers in this region are forced to drill deep wells to find irrigation water.
4. Crops in this region were damaged by unfavorable growing conditions.
5. The instrument will measure typical operating conditions.
6. After the funnel cloud passed, we found superficial damage.
7. Just before a dangerous storm, the sky takes on an unusual hue.
8. Conditions in the passage may be dangerous for small craft.
9. This is a state-of-the-art marine transmitter.
10. He went to the store for an electrical acoustic transducer.
11. Every family should have a sphygmomanometer to help monitor family health.

12. An anchor is a device used to hold a boat in one place.
13. The two parts are placed close to one another.
14. The working conditions in the basement were unbelievably poor.
15. XXX is a unique city.

EXERCISE

Directions: Assume a novice audience as you write answers for the following problems.

1. Make *micron* an understandable unit of measurement by comparing it to something the reader knows. (Look up *micron* in a dictionary, if you do not know what it means.)
2. Use an analogy or comparison to explain the strength of a 8.5 (Richter Scale) earthquake.

CLARITY

Good technical prose avoids wordiness, double meanings, and language offensive to some readers. This section explains how to avoid problems of wordiness, unnecessary jargon, careless modifiers and pronouns, and sexual bias in word choice.

Wordiness

Some writers mistake big words and long sentences for good writing. To show how educated they are, they stack word upon word. The result—if it is ever read—is confusing and unpleasant. Good writing is emphatic economical, and, above all else, *clear*. The following example shows the difference:

Original: Weather is always present and, to a varying degree, is a continuing factor in the conduct of any military action. Its effects are so numerous and widely varied that they can be properly evaluated only when treated within their relationship to all other factors contributing to the manner of conducting specific operations. This becomes particularly manifest in the application of weather forcasts to planned or contemplated operations. Any operation, therefore, the conduct of which might be affected by weather conditions, demands a proper consideration of a reliable forecast to cover the action and appropriate adjustments in plans necessitated by the weather forecast if effective conduct is to be assured.

Revision: When you plan a military operation, consult a weather forecast.

This example and revision are from *Guide for Air Force writing*, 1973.

As you revise your own writing, cut wordiness wherever you find it—but never at the expense of clarity.

One way to reduce wordiness is to prefer short words and clear explanations. Compare the following lists:

PREFERRED FORM

utilize	use
perambulate	walk
discontinue	stop
negative	no
primary	first
compensate	pay
demonstrate	show
afford an opportunity	allow
at all times	always
despite the fact that	although
due to the fact that	because
costs the sum of	costs

The list could go on and on. Cutting wordiness involves nearly all the revisions mentioned in this chapter. This section points out one more aspect of the problem.

EXERCISE

Directions: Cut wordiness from the following sentences.

1. Despite the fact that the goods were marked "seconds," the client claims he wasn't properly informed.
2. Temperature is a most important factor in determining the ecological optimum and limits of crop growth.
3. He has been told to cease and desist from the illegal practice of hauling gravel from county land.
4. We will compensate you for the negative impact of the damaged goods on seasonal sales.
5. It is felt that in the future, salespeople, during their basic training period, should have indoctrination lectures on all aspects of the company.
6. I wish to take this opportunity to express my appreciation for your assistance.
7. This approach will help us in the promulgation and implementation of changes commensurate with company needs.

8. The environmental effects are so subtle and confounded with other effects that we neither realize nor appreciate them fully.
9. It is requested that prints of all appurtenances and facilities be forwarded to this office as soon as practical.
10. If we utilize our reserves to the optimum, discontinuity in our supply of parts, on a temporary basis, won't unduly impact our operation.

Jargon

Jargon, the specialized vocabulary and language of a particular group or profession, is a handy form of shorthand for people who recognize it. For outsiders, technical terms and acronyms produce confusion and frustration. An *acronym* is a word created from the first letters of other words—for example NATO means North Atlantic Treaty Organization.

Audience analysis (see Chapter 3) is your key to word selection. Decide upon the lowest level of technical comprehension you can expect in the readers of the memo, letter, or report you are writing. Then choose words that will communicate with that audience.

Study this example:

Example: Provide parallel BCD output, 42 bits, TTL compatible for driving remote displays, along with IRIG B formatted time.

Can you understand it? The high incidence of technical jargon, combined with three acronyms—*BCD, TTL, IRIG B*—makes this sentence difficult reading for a nonspecialist.

When you use acronyms for the first time, define them:

Positive Example: As you know, we are cooperating closely with you and the Northeastern Utilities Consortium (NEUC) to provide an Emergency and Process Information Computer system (EPIC).

Rules for Acronyms

1. Write the acronym in capital letters but do not use periods.
2. Avoid crowding single sentences and paragraphs with acronyms.
3. If several pages separate your original explanation of an acronym and its next occurrence, explain it again.

EXERCISE

Directions: Underline jargon or acronyms that a general reader probably would not understand.

1. The equipment data base is the primary resource upon which the system is organized.
2. This maintenance program is centered around CM and PM.
3. The most outstanding feature of the software is that it is menu-driven.
4. Corrective maintenance is initiated into the system by completing the necessary fields on the work request data entry screen.
5. The receiver features a BFO to help tune CW, SSB, and DSB signals.
6. The nurse told me the patient has GI problems.
7. To strengthen the seam in the garment, she used a stitch-in-a-ditch.
8. Good writing rarely uses the subjunctive mood.
9. Start with 80 grit and finish with 300.
10. He's suffering from hypertension.

EXERCISE

Directions: Analyze the use of acronyms in the following paragraph:

The team leader and NSP personnel shall finalize an approved format and content for the "standard" EPG, after which all EPGs shall be written and submitted to NSP for review in an order established by the team leader with NSP approval. The EPGs will be finalized as they are received from NSP.

1. What is the effect of the acronyms on general readability?
2. Would you be more comfortable as a reader if you knew EPG meant Emergency Procedure Guideline, and that NSP meant Northern States Power?
3. What is the basic rule writers should follow when including acronyms in their writing?

Careless Modifiers

A *modifier* is a word or group of words that describes or limits the meaning of another word. Modifiers help make writing specific and concrete. For example, the noun *transmitter* becomes more specific if *multi-band* or *mono-band* is placed before it, creating the term *multi-band transmitter*.

However, modifiers can also be a source of confusion, particularly if they are not placed close to the word modified. This section discusses dangling and squinting modifiers.

DANGLING MODIFIERS

A *dangling modifier* either appears to modify the wrong word or apparently has nothing to modify. It usually appears at the beginning of a sentence. Study this example:

Original: Driving down the road, the house came into view.

Driving down the road appears to modify *house*, but this is illogical because a house *cannot drive down a road*.

Revision: Driving down the road, I saw the house come into view.

I, the subject of the revised sentence, is capable of driving. Adding a subject makes the sentence logical. *Driving down the road* modifies *I*—and this is certainly better than having houses drive down roads!

To correct a dangling modifier, place a noun within the modifying phrase or in the slot immediately following it. Consider another example:

Original: Operating the equipment at optimum speed, the goal was easily reached in the first quarter.

Revision: Operating the equipment at optimum speed, company employees easily reached the goal in the first quarter.

The original version has *operating the equipment at optimum speed* modifying *the goal*. This does not make sense—a goal can't operate equipment; people do. The revision has the same phrase modifying *company employees*—someone capable of operating the equipment. In addition, the verb is now active. The original dangles; the revision doesn't.

Dangling modifiers are extremely common among business and technical writers who favor passive verbs (for a full discussion of the passive, review pp 253–258). Passive verbs emphasize the *what* of an action rather than the who. Study one more example:

Original: After two years of on-site work, the job was finished.

—*was finished* is passive.
—The *doer* of the action—is missing.
—*After two years of on-site work* is a dangling modifier.

Revision: After two years of on-site work, *Acme* finished the job.

—*Acme* is the *doer*.
—*finished* is now an active verb.
—*After two years of on-site work* now modifies *Acme* and no longer dangles.

Remember, a modifier dangles if it is not clearly (and closely) tied to the word it modifies.

EXERCISE

Directions: Find and correct the dangling modifiers in the following sentences (two sentences are correct). If necessary, add words to solve the problem.

1. Recognizing the problem of parts logistics, a plan was developed to avoid any shortages.
2. Concerned about possible voting irregularities, county officials were contacted by the committee.
3. After singing the national anthem, the President was given a plaque honoring his contribution.
4. Confused by the instruction manual, the part was damaged beyond repair.
5. To complete the job on schedule, the equipment must be in good repair.
6. By working overtime, the crew finished the road only one week behind schedule.
7. By careful management of resources, the job was brought in below bid.
8. Before installing the new computer equipment, the room must be cleared and rewired.
9. Rigged for silent running, the crew took the submarine down.
10. Putting out 5,000 copies a day, the copy machine lasted just six months.

SQUINTING MODIFIERS

A *squinting modifier* modifies either the words that come before or after it. Because the reader cannot tell which meaning is meant, confusion results. Consider this example:

Original: Supervisors remind employees regularly to wear hard hats.

Revision: Supervisors regularly remind employees to wear hard hats.

The original is confusing because *regularly* is a squinting modifier: you cannot tell whether it modifies the words before or after it. This kind of double meaning is not acceptable. The revision shifts *regularly* to a position adjacent to *remind*, leaving only one possible meaning.

Some modifiers create double meaning when they are separated from the word modified. Here is a simple example:

Original: The equipment was sold to a small company which later went bankrupt for $200,000.

Revision: The equipment was sold for $200,000 to a small company which later went bankrupt.

In the original sentence, *for $200,000* seems to modify *bankrupt* but could also modify *sold.* Moving the phrase next to *sold* solves the problem.

The basic strategy for eliminating squinting modifiers and misplaced modifiers is to move them to a location where only one meaning is possible. If necessary, rewrite the sentence.

EXERCISE

Directions: Eliminate confusion by moving squinting or ambiguous modifiers to a location where only one meaning is possible.

1. The company tried to hire him for two years.
2. *Acme* purchased the equipment from a local supplier who became the sole source for $200,000.
3. The BLM approved the permit to construct the drill mud storage reservoir on April 1, 1986.
4. The governor (device to limit speed) was damaged only when the engine overheated.
5. Overextending financial resources too frequently results in cash-flow problems.
6. The investigators presented facts only at the meeting.
7. Finally, locate the tank on the blueprint that is made of steel.
8. We agreed on the first of the month to make the changes.
9. The part arrived only after six inquiries.
10. He overcame the odds against success under the supervision of a corporate vice-president.

Careless Pronouns

A *pronoun* is a part of speech that substitutes for a noun. Ambiguity results if the connection between the pronoun and the noun it substitutes for (called its antecedent) is not clear.

Original: *Acme* made the sale and delivered the software package; it was a significant one.

The reader may be confused by the vague pronoun reference: Does *it* refer to the sale or to the package?

Revision: *Acme* made a significant sale and delivered the software package.

The revision makes it clear that the *sale* was the reference or antecedent of the pronoun *it*.

The following list contains most common pronouns:

- I, you, he, she, it, we, you, they, me, him, her, us, them
- this, that, these, those, who, whose, whom, which
- each, another, either, neither, many few, some, any

Using this list, you should not have much trouble identifying pronouns. To see if you are ready to correct problems in pronoun reference, study the next sample:

Original: The manager first met him when he was in California.

    —Whom does *he* refer to?

    —Is the reference of the pronoun clear?

Revision: When he was in California, the manager met Joe for the first time.

    —In the reversed sentence order, *When he was in California,* clearly refers to the manager. However, the pronoun *him* is still ambiguous until its reference *Joe* is included.

Pronouns enable you to avoid repeating a noun over and over again. However, good usage requires one clear referent—a noun—for each pronoun.

EXERCISE

Directions: Eliminate pronoun-antecedent confusion and vague pronoun reference in the following sentences.

1. It is imperative to rush through the overhaul request if we are going with it on March 1.
2. When this unit was designed, they hadn't even though of rules concerning the working environment.
3. When the inspector criticized Roy in public, he was extremely upset.
4. There is water in this diesel fuel; do you want to use it?
5. Purchasing has the traction paper and ribbons, but it isn't the kind we ordered.
6. It says in the maintenance manual that the cover is removable.
7. Each committee member must cast their vote before the end of the week.
8. We have improved our profit picture and increased cash flow, which worries our competitors.
9. Inadequate supervision has resulted in poor work; this is certainly a serious problem on this project.
10. Bob received a call from him when he was in Europe.

Sexual Bias

The Civil Rights Act of 1964, which outlawed discrimination in employment on grounds of race, color, religion, sex, or national origin (later extended to include age and handicapped status), and subsequent acts have made many aware of sexual bias in language. Business, science, and industry today are concerned with positive approaches to fairness doctrines of all kinds. This section defines sexual bias in language, discusses examples, lists common sexist terms and nonsexist alternatives, explains methods for avoiding sexism, and provides material for revision.

Sexist language is language that treats one sex differently from the other. It is words and phrases implying a sexual bias or stereotype. The most easily recognized form uses masculine words that supposedly refer to both sexes.

COMMON FORM	PREFERRED FORM
chairman	presiding officer, chair
salesman	sales representative
fireman	fire fighter
foreman	supervisor, group leader
mailman	carrier, delivery person
businessman	business executive
bakeryman	baker, pastry chef
statesman	leader, opinion maker
Congressman	representative, member of Congress
draftsman	draftsperson
man on the street	person on the street

When you are searching for the proper word, eliminating *man* whenever it is suffix is often the simplest thing to do. Such changes are easy—once you are sensitive to the problem and aware of alternatives. Women work at a wide range of jobs; many prefer not to be defined by the name of the other sex.

Similar objections are raised to words that unnecessarily identify a person as a woman (or a man):

COMMON FORM	PREFERRED FORM
policewoman	police officer
co-ed	student
lady plumber	plumber
housewife	homemaker
laundry lady	laundry worker
poetess	poet, author
farmerette	farmer
career girl	(name profession) financial planner, artist, etc.
stewardess	flight attendant

Probably the most common problem concerns dealing with pronouns. Because there is no neutral pronoun representing both sexes, some writers avoid singular pronouns or use a *he or she* combination. Study these examples:

Original: The technician must monitor temperature fluctuations hourly. After the first twelve hours of operation he. . . .

—To avoid saying *technician* over and over again in a document, many writers use a pronoun such as *he*.

—*Are all technicians male?*

Revision: Technicians must monitor temperature fluctuations hourly. After the first twelve hours of operation they. . . .

—The revision avoids the masculine pronoun *he* by making *technician* plural. The pronoun then becomes *they*, avoiding any assumption about the technician's gender.

Original: Each employee must make up his or her mind about how the increased premium will affect him or her.

—The double pronouns are awkward when repeated.

Revision: Employees must decide how the increased premium will affect them.

or

Each employee must decide how the increased premium will affect him or her.

—The first revision reads better than the second which, by eliminating one repetition of *his or her*, is awkward—though better than the original.

Guidelines for Using Pronouns

1. Do not use *she* for roles that have traditionally been female but actually are mixed: secretary, flight attendant, teacher, nurse, and so on. Such a use implies that these are female roles.
2. Use *he or she* sparingly to indicate an indefinite person:

   When a technician is hired, he or she must be informed about the new policy.

3. Use the plural whenever your meaning allows.

   Not: A flight attendant must detail her role in each incident.

   But: Flight attendants must detail their role in each incident.

4. Drop the pronoun altogether if it is not needed.

   Not: Every secretary needs his or her desk copy of basic reference books.

   But: Every secretary needs a desk copy of basic reference books.

Sexual stereotyping represents another aspect of sexism in language. A *stereotype* is a standardized mental picture held by members of a group. To assume, for example, that all nurses are female is sexual stereotyping. Study this example:

Original: When a businessman selects a secretary, he should choose her with care.

—This example unfairly assumes that all who operate businesses are men and all secretaries are women. This is sexual stereotyping.

Revision: A business executive should select a secretary with care.

—The revision eliminates *he* and *her*, making the sentence gender-neutral.

Some writers attempt to solve the problem of sexist language with a disclaimer at the beginning of a document:

Example: Although the words *he*, *him*, and *his* are used sparingly in this manual to enhance communication, they are not intended to be gender-driven nor to affront or discriminate against anyone reading *Construction Mechanic*.

Although showing awareness of the problem, such an approach is unsatisfactory to some readers.

There is no simple solution to all aspects of sexism in language. Careful writers are aware of potential problems and avoid offending readers.

EXERCISE

Directions: Revise the following sentences to eliminate sexist language and awkward constructions.

1. ~~Each~~ supervisor~~s~~ should notify ~~his~~ *all* crews of the proposed changes in work rules.
2. When ~~a doctor hires~~ *hiring* a new nurse, ~~he~~ *doctor's* should check ~~her~~ *the applicant's* résumé for evidence of sound technical competence as well as worthwhile experience. *should determine*
3. Each supervisor ~~must decide for himself or herself whether he or she is willing to spend his or her time~~ working on this problem. *Whether time should by spent*
4. ~~The~~ typical engineer~~s~~ enjoys ~~his~~ *their* work.
5. Traditionally, ~~men~~ *males* are strong, decisive, and athletic; ~~women~~ *Females* are sensitive, shy, and nonassertive.
6. Call your realtor and ~~give him~~ more incentive to sell.
7. (Letter sent in mass mailing): "Dear Housewife,"

8. (Office setting): I'll have my girl fire a memo off to you immediately.
9. An executive secretary has her hands full when the board meets.
10. He's working as a male nurse at Madison Hospital.

CORRECTNESS

Good business and technical writing, besides being economical, emphatic, concrete, and clear, is also grammatically correct. It conforms to standard practices in structure and punctuation. This section identifies four common problems of correctness and tells you how to avoid them:

1. Make subject and verb Agree
2. Eliminate sentence fragments
3. Punctuate sentences correctly
   —Comma-faults
   —run-on sentences
4. Make equal elements parallel

Make Subject and Verb Agree

A *sentence* is a group of words containing a subject and verb. The *subject* is a word or group of words about which the sentence makes a statement. It tells the topic of the sentence. It may appear anywhere in a sentence but usually appears at the beginning. A *verb* shows action or being and completes the thought of the sentence. Consider this simple example:

Example: The equipment is here.

—*equipment* is the subject
—*is* is the verb

A common error occurs when the subject and the verb do not agree in number. *Number* means singular or plural. If the subject is singular, then the verb must agree with it; that is, it must be singular also. Study this simple example:

Original: The equipment and material is here.

—*equipment* plus *material* make a plural subject
—*is* is singular
—The subject and verb *do not agree.*

Revision: The equipment and material are here.

As sentences become longer, words, clauses, and phrases often separate the subject and verb, making it difficult to identify the correct subject. Probably the most confusing situation results when one or more prepositional phrases separate the subject and verb.

	s. Prepositional Phrase v.

Original: A display (of hardwood and woodworking tools) are in the foyer.

—*of hardwood and woodworking tools* is a prepositional phrase.
—*display* (singular) is the subject
—*are* (plural) is the verb
—The subject and verb do not agree!

Revision: A display of hardwood and woodworking tools is in the foyer.

—*display* is the subject (singular)
—*is* is the verb (singular)
—The subject and verb now agree.

Prepositional phrases are modifiers; they *never* have a subject or verb in them. To avoid confusing prepositional phrases with the subject, read a sentence without the prepositional phrase(s): the subject and verb verb will usually be obvious. For example, very few writers would say *The display are here.*

Sometimes you may recognize the subject and not know whether it is singular or plural. *Collective nouns,* words referring to groups rather than individuals, and several other forms are confusing:

1. Collective nouns: crew, team, family, choir, group, army, class, crowd, administration, and so on.

   Example: The audience is ready for the performance.

   Although a collective noun has a singular form and refers to a group of members, consider it singular and make the verb singular too.

2. Either . . . or: Two subjects joined by *or, either . . . or,* or *neither . . . nor* ask the reader to choose between two things. The verb agrees with the subject nearer to it.

   Example: Neither the participants nor the leader is required to attend.

   Participants and *leader* are both subjects; however, the one nearer the verb is *leader,* which is singular. It agrees with *is,* which is also singular.

   If the two subjects were reversed, the verb would have to be a plural *are* to agree with *participants.*

   Example: Neither the leader nor the participants are required to attend.

3. Each and every: If *each* or *every* comes before the subject, then it is always singular, even if it contains several parts.

   Example: Every crew member and leader in the department was embarrassed by the incident.

4. Indefinite pronouns: each, either, everybody, everyone, everything, neither, nobody, none, no one.

   These words are generally considered singular.

   Example: <u>Each</u> of the following individuals <u>is</u> entitled to a refund.

5. There: When a sentence begins with the word *there*, look for the real subject after the verb.

   Example: There <u>are two</u> of them.

EXERCISE

Directions: Correct errors in subject/verb agreement.

1. A style number is a set of digits printed on clothing which indicate variations in material, design, and cut.
2. Each one of them have been carefully verified.
3. The administration have selected June 4th as the target completion day.
4. The new laboratory, along with the present facility, are more than adequate for our needs.
5. A major part of the test module experimental measurements are made in static or in-flight tests.
6. If the position of the flaps on the aircraft's wings are changed radically during a high-speed test, the unusual stress could cause serious damage.
7. Neither the plane nor the pilot are due to arrive until 1500.
8. None of the ten units tested meet our requirements for mpg and initial unit cost.
9. There is three versions that must be considered before either of the groups come to a decision.
10. The crew are an important part of the total effort.

Eliminate Sentence Fragments

A sentence must have a subject, a verb, and express a complete thought. A *sentence fragment* is a piece of a sentence posing as a complete thought. Because it is incomplete, the fragment may confuse the reader.

Length is not important as long as the sentence has a subject and verb and makes sense:

Example: <u>He finished.</u>

Sentence fragments often result when a writer cuts a modifying phrase from the main sentence. Study the example.

Original: When the shipment arrived, it was damaged. <u>Ripped and crushed almost beyond recognition.</u>

The phrase *ripped and crushed almost beyond recognition* does not make sense alone and has neither a subject nor a verb. To eliminate the fragment, hook it to the main sentence where it belongs:

Revision: When the shipment arrived it was damaged, ripped and crushed almost beyond recognition.

Sentence fragments also result when a modifying clause is separated from the main sentence. A *modifying clause* is a group of words containing a subject and verb but that does not make sense by itself.

Modifying Clause: Because he was ill

The example does not make sense. Alone it is a fragment. You wonder "What happened because he was ill"? If the modifying clause is connected to the main sentence, the combination—a complete sentence—is much easier to understand.

Revision: sentence modifying clause
He left work early because he was ill.

or

Because he was ill, he left work early.

Recognizing a modifying clause used as a fragment is easy because of the words that introduce them. There are two common sets:

Relative Pronouns: who, whom, which, that

Subordinate Conjunctions:

after	because	than
although	before	where
as	as if	if
since	when	while

as long as
as soon as
though
in order that
provided (that)
so (that)
whenever
wherever
unless

Compare the following forms:

Sentence: He went to town.
Fragment: *Because* he went to town.

Addition of *because* makes *He went to town* a fragment because it no longer expresses a complete thought: you wonder what happened.

Revision: <u>He missed the phone call</u> because he went to town.
               main clause or sentence

Sentence fragments generally are modifying phrases or clauses separated from a main sentence—usually the one immediately before the fragment. To correct the problem, hook the fragment to the appropriate sentence or add words to make it complete.

EXERCISE

Directions: Correct all fragments in the following sentences.

1. We investigated the possibility of outright purchase. While the other group studied leasing equipment.
2. He stood still. As if the sight of the burning truck was beyond his comprehension.
3. The design avoided combustible materials. The highest expected temperature being 350°F.
4. If the project comes in on time with no cost overruns.
5. The new manager instituted several changes. Many of which are impractical.
6. We are investigating IBM, Apple, and Tandy microcomputers. All excellent equipment for our present and future requirements as we now know them.
7. We've investigated thirty-one anomalies this year. The most we've had in ten years of operation.
8. The job completed after two years of construction and the expenditure of $7 million.
9. Provided that the research and development can be completed on schedule and provided a customer is available.
10. The boat is located in Seattle. Where the temperature rarely reaches freezing and the sailing season lasts up to eight months.

Punctuate Sentences Correctly

Two of the most common errors in sentence punctuation are the *comma fault* and *run-on sentence.*

Normally two sentences (main clauses) are connected by a comma and a joining word such as *and.* If only a comma is used, the resulting error is a *comma fault* or *comma splice.* If nothing is used to join them, the resulting error is a *run-on sentence.* Study the following examples:

Comma-Fault: The wind blew for days on end, snowdrifts ten–fifteen feet tall blocked the roads.

Run-on Sentence: The wind blew for days on end snowdrifts ten–fifteen feet tall blocked the roads.

Revisions: A. The wind blew for days on end, and snowdrifts ten–fifteen feet tall blocked the roads.

or

B. The wind blew for days on end; snowdrifts ten–fifteen feet tall blocked the roads.

Depending on which seems to read best in a given document, there are four basic options for correcting comma faults and run-on sentences.

OPTION 1

comma*   plus

- coordinate conjunction
- and
- but
- or
- nor
- yet—when it means *but*
- so—when it means *therefore*
- for—when it means *because*

Example: Parts are scarce, and they are expensive when you can find them.

Note: The comma is often omitted if the two sentences are short. However, it is never wrong to use a comma, even between short clauses.

OPTION 2

Connect the sentences with a semicolon.

Example: Parts are scarce; they are expensive when you can find them.

Although no connecting word is required with a semicolon, sometimes it is convenient to use one. When you use one of the following words (conjunctive adverbs) to join two sentences, put a semicolon before it and a comma after it.

Conjunctive Adverbs

for example	that is	besides
after all	also	second
furthermore	moreover	in contrast

similarly	however	still
nevertheless	on the other hand	perhaps
indeed	in fact	consequently
possibly	as a result	thus
hence	therefore	in other words
frankly	unfortunately	finally
in summary	to conclude	

This list is not complete, but it does include the most commonly used conjunctive adverbs.

Example: Parts are scare; therefore, they are expensive when you can find them.

OPTION 3

Make two separate sentences, particularly if the combination is long and wordy. (Be careful not to create two short, weak sentences.)

Example: Parts are scarce. They are expensive when you can find them.

OPTION 4

Subordinate one sentence to the other. Use a subordinate conjunction to introduce one clause. If the clause comes first, it is followed by a comma. If it comes last, no punctuation is required.

Example: Because parts are scarce, they are expensive when you find them.

or

Parts are expensive when you find them because they are scarce.

All four options are correct and useful. The basis for choosing one over another to correct a comma fault or run-on sentence is tied to variables such as sentence flow, purpose, and meaning.

EXERCISE

Directions: Using any of the four options, correct comma faults and run-on sentences.

1. The initial planning activity is die size estimation this is essentially a two-step process.
2. The branch office is in Richmond, VA, we made airline reservations for Tuesday.

3. The new computer does an excellent job, it will certainly help speed up our monthly reports.
4. The guard listened carefully, someone was coming down the hall.
5. The first version has several advantages, however, I favor the second.
6. Where has Phil Jones been he hasn't reported to work for two weeks.
7. He's extremely familiar with the control panel, he could operate the machine in total darkness.
8. The two are equal in quality but Item A is .20 cents lower in unit cost therefore it is the better choice.
9. Because management is seeking decentralization, this is an opportunity to innovate and show your ability.
10. We've experienced excessive downtime with our operating equipment adequate maintenance is difficult on double shifts.
11. Because of lower prices sales are up but net income is down.
12. There are 20 slabs measuring 15 feet 6 inches in length, each is 8 feet wide and 18 inches deep.
13. The road was blocked by a snowslide resort owners offered reduced rates to trapped skiers.
14. Cash flow problems have bothered the firm since it opened its doors as a result company officials have filed for protection under bankruptcy laws.
15. This is my final decision I'm not likely to change my mind.

Make Equal Elements Parallel

Parallelism, using the same grammatical structure for all items in a series or list, helps hold sentences together and provides a smooth flow to writing. *Faulty parallelism* occurs when ideas serving the same purpose do not have the same grammatical form. Consider several examples:

Faulty Parallelism: Your solution is inexpensive, slow, and it is difficult to implement.

Analysis:
‖ inexpensive = adjective
‖ slow = adjective
‖ it is difficult to implement = clause

To be parallel, all three elements in the series statement must be adjectives.

Revision: Your solution is inexpensive, slow, and difficult to implement.

Analysis: Your solution is—

‖ *inexpensive,* = adjective
‖ *slow* = adjective
and ‖ *difficult* = adjective
to implement.

The first element in the series establishes the grammatical pattern; all others must match it.

Faulty Parallelism: He <u>checked</u> the data, <u>evaluated</u> all aspects of the system, and <u>he consented</u> to a trial demonstration.

Analysis: ‖ *checked* = verb
              evaluated = verb
              he consented = clause

To be parallel, each main item in the series must be identical grammatically.

Revision: He <u>checked</u> the data, <u>evaluated</u> all aspects of the system, and <u>consented</u> to a trial demonstration.

Analysis: In the revision, three verbs are parallel: checked, evaluated, and consented.

Note: Faulty parallelism often occurs in items connected with the coordinate conjunctions *and*, *but*, *or*, and *nor*.

Faulty Parallelism: DDT <u>is found</u> not only in the soil, but it <u>is</u> also in plants, animals, and even humans.

Analysis: The first sentence has a passive verb: *is found.* The second sentence has an active verb: *is.* To be parallel or equal grammatically, both sentences must have the same type of verb.

Revision: DDT <u>is found</u> not only in the soil, but it <u>is</u> also <u>found</u> in plants, animals, and even humans.

or

DDT is found not only in the soil but also in plants, animals, and even humans.

The second revision reads much better than the first, although both are parallel. The second, instead of paralleling sentences, reduces one sentence to a phrase: *in the soil* is paralleled to *in plants, animals, and even humans*—two prepositional phrases.

Words, clauses, and phrases included in lists are considered equal in function and meaning and, therefore, must be grammatically equal in form also. Consider the following example:

Faulty Parallelism: 1. In<u>sp</u>ect the area to be trimmed.
                        2. <u>Allow</u> the tip of the line to do the cutting.

v.
3. <u>Allow</u> a side-to-side movement.

noun
4. <u>Grass</u> in excess of eight inches tall should be cut by working from top to bottom.

Analysis: Steps 1, 2, and 3 each begin with a verb and are, therefore, parallel. Step four begins with a noun and has a different form of verb.

Revision of   4. <u>Cut</u> grass in excess of eight inches by working from
Step 4:   top to bottom.

Step 4 now begins with a verb, making it parallel to the other three steps.

Remember, parallelism involves matching equal grammatical items—words, clauses, phrases, and even sentences. As you do the exercises, look for problems near the coordinate conjunctions (and, but, or, nor).

EXERCISE

Directions: Correct any errors in parallelism.

1. I favor changing managers for two reasons: to show people that we're concerned about problems and because we need a new direction.
2. The equipment was repaired in twenty-four hours, and it is now on the job.
3. He is an excellent manager, who tries to get projects out of the shop on time or getting more help to solve problems quickly.
4. We must improve in three important areas: marketing, keeping records, and in reaching consumers who need our product.
5. The following problems are associated with the production facility: improving the performance of employees through retraining, reduction of rejects because of contamination, improved safety habits, and elimination of communication problems.
6. Jones is the type of woman needed for the job and who knows that needs to be done.
7. The paint is peeling, it must have a new roof, the foundation is sinking, and the windows are broken. (If you cannot make all related elements in a series parallel, break the sentence into two clauses.)
8. We will fight them on the ocean, on the beaches, in the streets, in our homes if necessary, but we will not give up.
9. Sailing is an interesting hobby and which is also exciting and challenging.
10. She is hard-working, thoughtful, and she rarely loses her temper.

INDEX